δ Delta

Focus: Division

Instruction Manual

by Steven P. Demme

1-888-854-MATH (6284)
www.MathUSee.com

Math·U·See

1-888-854-MATH (6284)
www.MathUSee.com
Copyright © 2009 by Steven P. Demme

δ Delta

SCOPE AND SEQUENCE
HOW TO USE MATH-U-SEE

SUPPORT AND RESOURCES

LESSON 1   Rectangles, Factors and Product

LESSON 2   Division by 1 and 2, Symbols for Division

LESSON 3   Division by 10, Third Symbol for Division

LESSON 4   Division by 5 and 3

LESSON 5   Parallel and Perpendicular Lines

LESSON 6   Division by 9

LESSON 7   Finding the Area of a Parallelogram

LESSON 8   Division by 6

LESSON 9   Finding the Area of a Triangle

LESSON 10   Division by 4

LESSON 11   Finding the Average

LESSON 12   Division by 7 and 8

LESSON 13   Finding the Area of a Trapezoid

LESSON 14   Thousands, Millions, and Place-Value Notation

LESSON 15   Billions, Trillions, and Expanded Notation

LESSON 16   Division by a Single Digit with Remainder

LESSON 17   Upside-Down Multiplication

LESSON 18   Division with Double-Digit Factors

LESSON 19   Division, Three Digit by One Digit

LESSON 20   Division, Three Digit by One Digit

LESSON 21   Rounding to 10, 100, 1,000, and Estimation

LESSON 22   Division, Three Digit by Two Digit

LESSON 23   Division, Four Digit by One Digit

LESSON 24   Division, Four Digit by Two Digit

LESSON 25   Division, Multiple Digit by Multiple Digit

LESSON 26   Volume

LESSON 27   Fraction of a Number

LESSON 28   Roman Numerals: I, V, X, L, and C

LESSON 29   Fraction of One

LESSON 30   Roman Numerals: D, M, and Multiples of 1,000

STUDENT SOLUTIONS
TEST SOLUTIONS

GLOSSARY OF TERMS
MASTER INDEX FOR GENERAL MATH
INDEX FOR DELTA

 Math·U·See

SCOPE & SEQUENCE

Math-U-See is a complete and comprehensive K-12 math curriculum. While each book focuses on a specific theme, Math-U-See continuously reviews and integrates topics and concepts presented in previous levels.

Primer

α Alpha | Focus: Single-Digit Addition and Subtraction

β Beta | Focus: Multiple-Digit Addition and Subtraction

γ Gamma | Focus: Multiplication

δ Delta | Focus: Division

ε Epsilon | Focus: Fractions

ζ Zeta | Focus: Decimals and Percents

Pre-Algebra

Algebra 1

Stewardship*

Geometry

Algebra 2

Pre Calculus With Trigonometry

*Stewardship is a biblical approach to personal finance. The requisite knowledge for this curriculum is a mastery of the four basic operations, as well as fractions, decimals, and percents. In the Math-U-See sequence these topics are thoroughly covered in Alpha through Zeta. We also recommend Pre-Algebra and Algebra 1 since over half of the lessons require some knowledge of algebra. Stewardship may be studied as a one-year math course or in conjunction with any of the secondary math levels.

Five Minutes for Success

Welcome to Delta. I believe you will have a positive experience with the unique Math-U-See approach to teaching math. These first few pages explain the essence of this methodology which has worked for thousands of students and teachers. I hope you will take five minutes and read through these steps carefully.

I am assuming your student has a thorough grasp of addition, subtraction, and multiplication.

If you are using the program properly and still need additional help, you may contact your authorized representative, or visit Math-U-See online at http://www.mathusee.com/support.html

— S. Demme

The Goal of Math-U-See

The underlying assumption or premise of Math-U-See is that the reason we study math is to apply math in everyday situations. Our goal is to help produce confident problem solvers who enjoy the study of math. These are students who learn their math facts, rules, and formulas and are able to use this knowledge to solve word problems and real life applications. Therefore, the study of math is much more than simply committing to memory a list of facts. It includes memorization, but it also encompasses learning the underlying concepts of math that are critical to successful problem solving.

More than Memorization

Many people confuse memorization with understanding. Once while I was teaching seven junior high students, I asked how many pieces they would each receive if there were fourteen pieces. The students' response was, "What do we do: add, subtract, multiply, or divide?" Knowing how to divide is important, understanding when to divide is equally important.

THE SUGGESTED 4-STEP MATH-U-SEE APPROACH

In order to train students to be confident problem solvers, here are the four steps that I suggest you use to get the most from the Math-U-See curriculum.

Step 1. Prepare for the Lesson
Step 2. Present the New Topic
Step 3. Practice for Mastery
Step 4. Progression after Mastery

Step 1. Prepare for the Lesson.

Watch the DVD to learn the new concept and see how to demonstrate this concept with the manipulatives when applicable. Study the written explanations and examples in the instruction manual. Many students watch the DVD along with their instructor.

Step 2. Present the New Topic

Now that you have studied the new topic choose problems from the first lesson practice page to present the new concept to your students.

a. **Build:** Use the manipulatives to demonstrate the problems from the worksheet.

b. **Write:** Record the step-by-step solutions on paper as you work them through with manipulatives.

c. **Say:** Explain the *why* and *what* of math as you build and write.

Do as many problems as you feel are necessary until the student is comfortable with the new material. One of the joys of teaching is hearing a student say *"Now I get it!"* or *"Now I see it!"*

Step 3. Practice for Mastery.

Using the examples and the lesson practice problems from the student text, have the students practice the new concept until they understand it. It is one thing for students to watch someone else do a problem, it is quite another to do the same problem themselves. Do enough examples together until they can do them without assistance.

Do as many of the lesson practice pages as necessary (not all pages may be needed) until the students remember the new material and gain understanding. Give special attention to the word problems, which are designed to apply the concept being taught in the lesson.

Another resource is the Math-U-See web site which has online drill and downloadable worksheets for more practice. Go to www.mathusee.com and select "Online Helps."

Step 4. Progression after Mastery.

Once mastery of the new concept is demonstrated, proceed to the systematic review pages for that lesson. Mastery can be demonstrated by having each student teach the new material back to you. The goal is not to fill in worksheets, but to be able to teach back what has been learned.

The systematic review worksheets review the new material as well as provide practice of the math concepts previously studied. Remediate missed problems as they arise to ensure continued mastery.

Proceed to the lesson tests. These were designed to be an assessment tool to help determine mastery, but they may also be used as extra worksheets. Your students will be ready for the next lesson only after demonstrating mastery of the new concept and continued mastery of concepts found in the systematic review worksheets.

Confucius was reputed to have said, "Tell me, I forget; Show me, I understand; Let me do it, I will remember." To which we add, **"Let me teach it and I will have achieved mastery!"**

Length of a Lesson

So how long should a lesson take? This will vary from student to student and from topic to topic. You may spend a day on a new topic, or you may spend several days. There are so many factors that influence this process that it is impossible to predict the length of time from one lesson to another. I have spent three days on a lesson and I have also invested three weeks in a lesson. This occurred in the same book with the same student. If you move from lesson to lesson too quickly without the student demonstrating mastery, he will become overwhelmed and discouraged as he is exposed to more new material without having learned the previous topics. But if you move too slowly, your student may become bored and lose interest in math. But I believe that as you regularly spend time working along with your student, you will sense when is the right time to take the lesson test and progress through the book.

By following the four steps outlined above, you will have a much greater opportunity to succeed. Math must be taught sequentially, as it builds line upon line and precept upon precept on previously learned material. I hope you will try this methodology and move at your student's pace. As you do, I think you will be helping to create a confident problem solver who enjoys the study of math.

ONGOING SUPPORT
AND ADDITIONAL RESOURCES

Welcome to the Math-U-See Family!

Now that you have invested in your children's education, I would like to tell you about the resources that are available to you. Allow me to introduce you to your regional representative, our ever improving website, the Math-U-See blog, our new free e-mail newsletter, the online Forum, and the Users Group.

Most of our regional **Representatives** have been with us for over 10 years. What makes them unique is their desire to serve and their expertise. They have all used Math-U-See and are able to answer most of your questions, place your student(s) in the appropriate level, and provide knowledgeable support throughout the school year. They are wonderful!

Come to your local curriculum fair where you can meet your rep face-to-face, see the latest products, attend a workshop, meet other MUS users at the booth, and be refreshed. We are at most curriculum fairs and events. To find the fair nearest you, click on "Events Calendar" under "News."

The **Website**, at www.mathusee.com, is continually being updated and improved.It has many excellent tools to enhance your teaching and provide more practice for your student(s).

ONLINE DRILL

Let your students review their math facts online. Just enter the facts you want to learn and start drilling. This is a great way to commit those facts to memory.

WORKSHEET GENERATOR

Create custom worksheets to print out and use with your students. It's easy to use and gives you the flexibility to focus on a specific lesson. Best of all — it's free!

Math-U-See Blog

Interesting insights and up-to-date information appear regularly on the Math-U-See Blog. The blog features updates, rep highlights, fun pictures, and stories from other users. Visit us and get the latest scoop on what is happening .

Email Newsletter

For the latest news and practical teaching tips, sign up online for the free Math-U-See e-mail newsletter. Each month you will receive an e-mail with a teaching tip from Steve as well as the latest news from the website. It's short, beneficial, and fun. Sign up today!

The Math-U-See Forum and the Users Group put the combined wisdom of several thousand of your peers with years of teaching experience at your disposal.

Online Forum

Have a question, a great idea, or just want to chitchat with other Math-U-See users? Go to the online forum. You can also use the forum to post a specific math question if you are having difficulty in a certain lesson. Head on over to the forum and join in the discussion.

Yahoo Users Group

The MUS-users group was started in 1998 for lovers and users of the Math-U-See program. It was founded by two home-educating mothers and users of Math-U-See. The backbone of information and support is provided by several thousand fellow MUS users.

For Specific Math Help

When you have watched the DVD instruction and read the instruction manual and still have a question, we are here to help. Call your local rep, click the support link and e-mail us here at the home office, or post your question on the forum. Our trained staff has used Math-U-See themselves and are available to answer a question or walk you through a specific lesson.

Feedback

Send us an e-mail by clicking the feedback link. We are here to serve you and help you teach math. Ask a question, leave a comment, or tell us how you and your student are doing with Math-U-See.

Our hope and prayer is that you and your students will be equipped to have a successful experience with math!

Blessings,

Steve

Steve Demme

Rectangles, Factors and Product
Solving for an Unknown

Rectangle means right angle. "Rect" comes from a German word that means right. A *right angle* is a square corner. If you find an object with four square corners, it is a rectangle. Look around you and see how many rectangles there are. This piece of paper is a rectangle. How many others can you identify?

A *square* is a special kind of rectangle. It has four right angles, so it is a rectangle. But it also has four sides that are the same length. So a rectangle with all four sides the same length is a square.

A rectangle is measured by its dimensions. A *dimension* is the length of a side.

Example 1

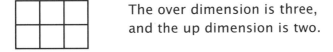

The over dimension is three, and the up dimension is two.

A rectangle also has area. You see that by the squares inside it. The dimensions tell how long the sides or edges are. The *area* tells how many squares are in the inside. In the rectangle above, that would be six.

Example 2

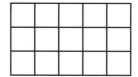

The over dimension is five,
and the up dimension is three.

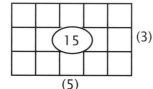

(3)   The area is 15 square units.

(5)

When we taught multiplication, we used the rectangle to illustrate a multiplication problem. We called the dimensions the *factors*, and the area was the *product*. Example 1 is a picture of 3 x 2 = 6 and example 2 is 5 x 3 = 15. When multiplying, we are given the factors and we have to find the product. In division, we will be given the product and one factor, and we'll be asked to find the missing factor.

This is just like solving for an unknown, which we did while learning our multiplication facts in the *Gamma* book.

Solving for an Unknown - How you verbalize the equation can be the key to understanding it. 2X = 12 can be read as "Two counted how many times is 12" or "What number counted two times is 12?" because of the ***commutative property*** of multiplication, which states that you can change the order of the factors. You might also verbalize this as "Two times what equals 12," or "What times two equals 12?" or "Two somethings equals 12." Any of these ways is acceptable. Choose whichever one makes it easier for the student to understand.

Example 3 illustrates this with the manipulative blocks. Study these examples until this important concept is understood.

Example 3

Solve for the unknown, or find out the value of "X" in 2X = 12.

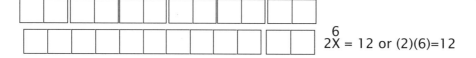

$$2X = 12 \text{ or } (2)(6)=12$$

I solved it as "How many 2s can I count out of 12?" or "What times 2 equals 12?" Using the blocks, you can see that there are six 2s in 12. So 2 x 6 = 12 and the missing factor is 6. When you find the answer, simply write a 6 just above the X. The student exercises may use other letters for the unknown instead of X.

WORD PROBLEM TIPS

Parents often find it challenging to teach children how to solve word problems. Here are some suggestions for helping your student learn this important skill.

The first step is to realize that word problems require both reading and math comprehension. Don't expect a child to be able to solve a word problem if he does not thoroughly understand the math concepts involved. On the other hand, a student may have a math skill level that is stronger than his or her reading comprehension skills. Below are a number of strategies to improve comprehension skills in the context of story problems. You may decide which ones work best for you and your child.

Strategies for word problems:

1. Ignore numbers at first and read the story. It may help some students to read the question aloud. Every word problem tells a story. Before deciding what math operation is required, let the student retell the story in his own words. Who is involved? Are they receiving gifts, losing something, or dividing a treat?

2. Relate the story to real life, perhaps by using names of family members. For some students, this makes the problem more interesting and relevant.

3. Build, draw, or act out the story. Use the blocks or actual objects when practical. Especially in the lower levels, you may require the student to use the blocks for word problems, even when the facts have been learned. Don't be afraid to use a little drama as well. The purpose is to make it as real and meaningful as possible.

4. Look for the common language used in a particular kind of problem. Pay close attention to the word problems on the lesson practice pages, as they model different kinds of language that may be used for the new concept just studied. For example, "altogether" indicates addition. These "key words" can be useful clues, but should not be substitutes for understanding.

5. Look for practical applications that use the concept and ask questions in that context.

6. Have the students invent word problems to illustrate their number problems from the lesson.

Cautions

1. Unneeded information may be included in the problem. For example, we may be told that Suzie is eight years old, but the eight is irrelevant when adding up the number of gifts she received.

2. Some problems may require more than one step to solve. Model these questions carefully.

3. There may be more than one way to solve some problems. Experience will help the student choose the easier or preferred method.

4. Estimation is a valuable tool for checking an answer. If an answer is unreasonable, it is possible that the wrong method was used to solve the problem.

Division by 1 and 2, Symbols for Division
Solving for an Unknown

There is an interactive math-facts practice page available online at mathusee.com/ drillsheet.html.

Once the student grasps that division is looking for the missing factor, the division facts should be a review of his knowledge of the multiplication facts. Please make sure that the multiplication facts are mastered before proceeding further in this book. The first items the student must become familiar with are the symbols for division. Two of the three symbols are introduced in this lesson. The first is a horizontal line as shown below, and the second is its derivation "÷." See figure 1.

Figure 1

$$\frac{6}{2}$$

The first symbol is a line between the numbers to be divided.

$6 \div 2$

Then the 6 was moved to the left of the line and the 2 to the right of the line, and dots were placed on the top and bottom of the line to remind us where the numbers used to be. Hence ÷ is the second symbol.

Another key to dividing is how one verbalizes this problem. The most common way is "Six divided by two." But to enhance understanding, consider "How many twos can I count out of six?" or "How many groups of two can I count out of six?" This is the same problem as $2X = 6$ that we have been doing in multiplication. There we read it as "Two times what number is the same as six?" That still applies here for a third option. Find out which verbalization turns on the light for the student and use that one.

In 2X = 6, notice that the product and one factor are given, and we have to find the missing factor. This is division, and solving for the unknown helps us to connect division and multiplication. In multiplication, we are given the two factors, and we need to find the product. In division we are given the product and one factor, and we need to find the missing factor. That is why division is the inverse of multiplication.

The division facts to be learned in this lesson are easier than most, so focus on the symbols and how to verbalize the problems as described in example 1.

Example 1

$$\frac{6}{2}$$

1. "What times two is equal to six?"
2. "Two times what is equal to six?"
3. "How many twos can I count out of six?"

$6 \div 2$

4. "Six divided by two equals what number?"

This is a good time to show all of the division facts on a chart. After the student learns several facts, color or circle the ones learned to encourage him in his progress. I will put a small chart in each of the lessons in the instruction manual and circle the facts currently being studied, as well as those already learned. There is a chart for the student after the lesson 2 worksheets in the student text.

$1 \div 1$	$2 \div 2$	$3 \div 3$	$4 \div 4$	$5 \div 5$	$6 \div 6$	$7 \div 7$	$8 \div 8$	$9 \div 9$	$10 \div 10$
$2 \div 1$	$4 \div 2$	$6 \div 3$	$8 \div 4$	$10 \div 5$	$12 \div 6$	$14 \div 7$	$16 \div 8$	$18 \div 9$	$20 \div 10$
$3 \div 1$	$6 \div 2$	$9 \div 3$	$12 \div 4$	$15 \div 5$	$18 \div 6$	$21 \div 7$	$24 \div 8$	$27 \div 9$	$30 \div 10$
$4 \div 1$	$8 \div 2$	$12 \div 3$	$16 \div 4$	$20 \div 5$	$24 \div 6$	$28 \div 7$	$32 \div 8$	$36 \div 9$	$40 \div 10$
$5 \div 1$	$10 \div 2$	$15 \div 3$	$20 \div 4$	$25 \div 5$	$30 \div 6$	$35 \div 7$	$40 \div 8$	$45 \div 9$	$50 \div 10$
$6 \div 1$	$12 \div 2$	$18 \div 3$	$24 \div 4$	$30 \div 5$	$36 \div 6$	$42 \div 7$	$48 \div 8$	$54 \div 9$	$60 \div 10$
$7 \div 1$	$14 \div 2$	$21 \div 3$	$28 \div 4$	$35 \div 5$	$42 \div 6$	$49 \div 7$	$56 \div 8$	$63 \div 9$	$70 \div 10$
$8 \div 1$	$16 \div 2$	$24 \div 3$	$32 \div 4$	$40 \div 5$	$48 \div 6$	$56 \div 7$	$64 \div 8$	$72 \div 9$	$80 \div 10$
$9 \div 1$	$18 \div 2$	$27 \div 3$	$36 \div 4$	$45 \div 5$	$54 \div 6$	$63 \div 7$	$72 \div 8$	$81 \div 9$	$90 \div 10$
$10 \div 1$	$20 \div 2$	$30 \div 3$	$40 \div 4$	$50 \div 5$	$60 \div 6$	$70 \div 7$	$80 \div 8$	$90 \div 9$	$100 \div 10$

Division Facts Sheet

$\div 1$	$\div 2$	$\div 3$	$\div 4$	$\div 5$	$\div 6$	$\div 7$	$\div 8$	$\div 9$	$\div 10$
$1 \div 1$	$2 \div 2$	$3 \div 3$	$4 \div 4$	$5 \div 5$	$6 \div 6$	$7 \div 7$	$8 \div 8$	$9 \div 9$	$10 \div 10$
$2 \div 1$	$4 \div 2$	$6 \div 3$	$8 \div 4$	$10 \div 5$	$12 \div 6$	$14 \div 7$	$16 \div 8$	$18 \div 9$	$20 \div 10$
$3 \div 1$	$6 \div 2$	$9 \div 3$	$12 \div 4$	$15 \div 5$	$18 \div 6$	$21 \div 7$	$24 \div 8$	$27 \div 9$	$30 \div 10$
$4 \div 1$	$8 \div 2$	$12 \div 3$	$16 \div 4$	$20 \div 5$	$24 \div 6$	$28 \div 7$	$32 \div 8$	$36 \div 9$	$40 \div 10$
$5 \div 1$	$10 \div 2$	$15 \div 3$	$20 \div 4$	$25 \div 5$	$30 \div 6$	$35 \div 7$	$40 \div 8$	$45 \div 9$	$50 \div 10$
$6 \div 1$	$12 \div 2$	$18 \div 3$	$24 \div 4$	$30 \div 5$	$36 \div 6$	$42 \div 7$	$48 \div 8$	$54 \div 9$	$60 \div 10$
$7 \div 1$	$14 \div 2$	$21 \div 3$	$28 \div 4$	$35 \div 5$	$42 \div 6$	$49 \div 7$	$56 \div 8$	$63 \div 9$	$70 \div 10$
$8 \div 1$	$16 \div 2$	$24 \div 3$	$32 \div 4$	$40 \div 5$	$48 \div 6$	$56 \div 7$	$64 \div 8$	$72 \div 9$	$80 \div 10$
$9 \div 1$	$18 \div 2$	$27 \div 3$	$36 \div 4$	$45 \div 5$	$54 \div 6$	$63 \div 7$	$72 \div 8$	$81 \div 9$	$90 \div 10$
$10 \div 1$	$20 \div 2$	$30 \div 3$	$40 \div 4$	$50 \div 5$	$60 \div 6$	$70 \div 7$	$80 \div 8$	$90 \div 9$	$100 \div 10$

Division by 10, Third Symbol for Division
for Division

Now that the one and two facts have been mastered, we will learn the ten facts and circle them on the chart on the next page. The third symbol of division is the half rectangle. Some call it a house or a box. It is very effective in assisting us in relating division, which we are now learning, to multiplication, which we have already learned. See figure 1.

Figure 1

$$10 \overline{\smash{)}\,40}^{\,?}$$

The product is 40 and the one factor is 10. We need to find the missing factor. Remember the rectangle reviewed in lesson 1, which we used extensively when teaching multiplication. The factors are the outside dimensions, and the product is the area.

$$10 \uparrow 40 \longrightarrow^{?}$$

When we built rectangles, we often referred to the up factor and the over factor, which is still a legitimate option if it proves helpful for the student.

$$\frac{40}{10} \qquad 40 \div 10$$

All of the problems in figure 1 are the same. They are just expressed with different symbolism. They may be verbalized several different ways.

1. "What times 10 is equal to 40?"
2. "10 times what is equal to 40?"
3. "How many 10s can I count out of 40?"
4. "40 divided by 10 equals what number?"

As shown by the rectangle in figure 2, the answer is four. Using the blocks, you can build all the 10 facts.

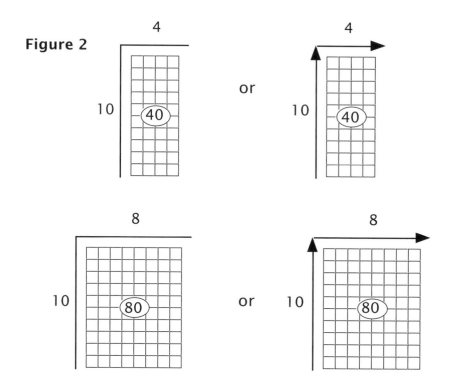

Figure 2

Division isn't commutative, but multiplication is. The problem 40 ÷ 10 is very different than the problem 10 ÷ 40. But every division problem may be expressed as a multiplication problem. In figure 3, the problem solved in figure 1 has the same product, but the factors have been changed. We see that 4 x 10 = 40 and 10 x 4 = 40. Both problems have the same product of 40. So you may change the factors and still produce the same answer. When you turn the rectangle, it still has the same factors and product or the identical dimensions and area.

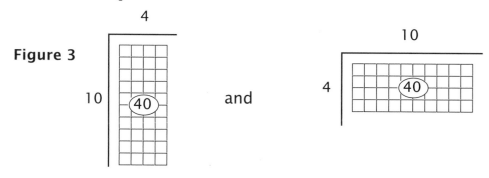

Figure 3

The grid below shows that when you learn the 10 facts on the far right, such as $10 \div 10 = 1$ and $20 \div 10 = 2$, you are also learning the facts on the bottom line that have 10 as a factor, such as $10 \div 1 = 10$ and $20 \div 2 = 10$.

$1 \div 1$	$2 \div 2$	$3 \div 3$	$4 \div 4$	$5 \div 5$	$6 \div 6$	$7 \div 7$	$8 \div 8$	$9 \div 9$	$10 \div 10$
$2 \div 1$	$4 \div 2$	$6 \div 3$	$8 \div 4$	$10 \div 5$	$12 \div 6$	$14 \div 7$	$16 \div 8$	$18 \div 9$	$20 \div 10$
$3 \div 1$	$6 \div 2$	$9 \div 3$	$12 \div 4$	$15 \div 5$	$18 \div 6$	$21 \div 7$	$24 \div 8$	$27 \div 9$	$30 \div 10$
$4 \div 1$	$8 \div 2$	$12 \div 3$	$16 \div 4$	$20 \div 5$	$24 \div 6$	$28 \div 7$	$32 \div 8$	$36 \div 9$	$40 \div 10$
$5 \div 1$	$10 \div 2$	$15 \div 3$	$20 \div 4$	$25 \div 5$	$30 \div 6$	$35 \div 7$	$40 \div 8$	$45 \div 9$	$50 \div 10$
$6 \div 1$	$12 \div 2$	$18 \div 3$	$24 \div 4$	$30 \div 5$	$36 \div 6$	$42 \div 7$	$48 \div 8$	$54 \div 9$	$60 \div 10$
$7 \div 1$	$14 \div 2$	$21 \div 3$	$28 \div 4$	$35 \div 5$	$42 \div 6$	$49 \div 7$	$56 \div 8$	$63 \div 9$	$70 \div 10$
$8 \div 1$	$16 \div 2$	$24 \div 3$	$32 \div 4$	$40 \div 5$	$48 \div 6$	$56 \div 7$	$64 \div 8$	$72 \div 9$	$80 \div 10$
$9 \div 1$	$18 \div 2$	$27 \div 3$	$36 \div 4$	$45 \div 5$	$54 \div 6$	$63 \div 7$	$72 \div 8$	$81 \div 9$	$90 \div 10$
$10 \div 1$	$20 \div 2$	$30 \div 3$	$40 \div 4$	$50 \div 5$	$60 \div 6$	$70 \div 7$	$80 \div 8$	$90 \div 9$	$100 \div 10$

Mastery of multiplication is essential for success in division. If you find that you need to review multiplication, consult the Math-U-See web site, which provides online drill and downloadable worksheets. Click on *www.mathusee.com* and look for "Online Helps."

Division by 5 and 3

We have seen from the first three lessons that division is built upon an understanding of multiplication.

Make sure the student has mastered the three and five multiplication facts before doing this lesson. If you need to take time to review these facts, please don't hesitate. Math is sequential and builds upon previously learned concepts. If you need to back up and learn multiplication, consider using the Math-U-See *Gamma* level.

When I was in school, we were taught division with the expression "How many threes goes into 27?" This phrase "goes into" degenerated into "gazinda." Even today, you will hear "How many threes gazinda 27?" Now if this reaches your ears, you will know its origin.

In a division problem there are names for the three components. They are *divisor*, *dividend*, and *quotient*. These are shown in figure 1. We don't use these a lot, but it is good to know them for the future. Notice in figure 2 how multiplication and division are related.

Figure 1
$$\text{divisor} \overline{)\text{dividend}}^{\text{quotient}}$$

Figure 2
$$\text{factor} \overline{)\text{product}}^{\text{factor}}$$

Learn all of the five facts and three facts. Take whatever time you need to master these before moving to the next lesson.

Example 1

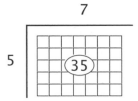

$$\frac{35}{5} = \qquad 35 \div 5 =$$

1. "What times five is equal to 35?"
2. "Five times what is equal to 35?"
3. "How many fives can I count out of 35?"
4. "35 divided by five equals what number?"

Example 2

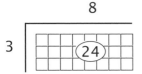

$$\frac{24}{3} = \qquad 24 \div 3 =$$

1. "What times three is equal to 24?"
2. "Three times what is equal to 24?"
3. "How many threes can I count out of 24?"
4. "24 divided by three equals what number?"

1 ÷ 1	2 ÷ 2	3 ÷ 3	4 ÷ 4	5 ÷ 5	6 ÷ 6	7 ÷ 7	8 ÷ 8	9 ÷ 9	10 ÷ 10
2 ÷ 1	4 ÷ 2	6 ÷ 3	8 ÷ 4	10 ÷ 5	12 ÷ 6	14 ÷ 7	16 ÷ 8	18 ÷ 9	20 ÷ 10
3 ÷ 1	6 ÷ 2	9 ÷ 3	12 ÷ 4	15 ÷ 5	18 ÷ 6	21 ÷ 7	24 ÷ 8	27 ÷ 9	30 ÷ 10
4 ÷ 1	8 ÷ 2	12 ÷ 3	16 ÷ 4	20 ÷ 5	24 ÷ 6	28 ÷ 7	32 ÷ 8	36 ÷ 9	40 ÷ 10
5 ÷ 1	10 ÷ 2	15 ÷ 3	20 ÷ 4	25 ÷ 5	30 ÷ 6	35 ÷ 7	40 ÷ 8	45 ÷ 9	50 ÷ 10
6 ÷ 1	12 ÷ 2	18 ÷ 3	24 ÷ 4	30 ÷ 5	36 ÷ 6	42 ÷ 7	48 ÷ 8	54 ÷ 9	60 ÷ 10
7 ÷ 1	14 ÷ 2	21 ÷ 3	28 ÷ 4	35 ÷ 5	42 ÷ 6	49 ÷ 7	56 ÷ 8	63 ÷ 9	70 ÷ 10
8 ÷ 1	16 ÷ 2	24 ÷ 3	32 ÷ 4	40 ÷ 5	48 ÷ 6	56 ÷ 7	64 ÷ 8	72 ÷ 9	80 ÷ 10
9 ÷ 1	18 ÷ 2	27 ÷ 3	36 ÷ 4	45 ÷ 5	54 ÷ 6	63 ÷ 7	72 ÷ 8	81 ÷ 9	90 ÷ 10
10 ÷ 1	20 ÷ 2	30 ÷ 3	40 ÷ 4	50 ÷ 5	60 ÷ 6	70 ÷ 7	80 ÷ 8	90 ÷ 9	100 ÷ 10

Parallel and Perpendicular Lines
Solving for an Unknown

Parallel Lines - Two straight lines that lie in the same plane (any flat surface, such as a piece of paper, the floor, or a desktop) and never cross or intersect are said to be **parallel**. Because they never touch, they are always the same distance apart. It is possible to have two lines in three-dimensional space that never intersect and are not parallel. If you take two lines represented by yardsticks, it is possible to position them in such a way that they never would intersect, even if the lines represented by them were extended at both ends. These are called skew lines.

But our discussion is with lines in the same plane that never touch. A line is shown with arrows at both ends to indicate that it goes on indefinitely in both directions. If it doesn't have the arrows, then it is only a segment of a line. Some examples of parallel lines are lines on a piece of paper or railroad tracks that are perfectly straight. Look around for other examples. When spelling parallel, notice that the two "ls" in the word are parallel to each other. Notice the pictures in figure 1. The symbol for parallel is ||.

Figure 1

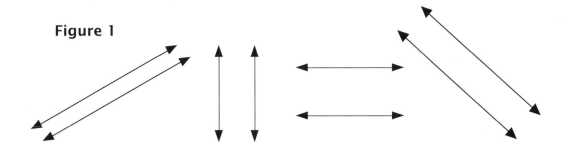

Perpendicular Lines - Two straight lines that lie in the same plane and do intersect and make a ***right angle***, or square corner, are said to be ***perpendicular lines***. To show that they form this right angle or square corner, we put a little square where they meet. This means the two lines are perpendicular. Two roads in a town can be perpendicular at an intersection if they form a right angle. A cross is made of two perpendicular lines. A kite is made up of two perpendicular line segments. Notice the pictures in figure 2. The symbol for perpendicular is ⊥.

Figure 2

LESSON 6

Division by 9

You can tell whether a number is divisible by nine by adding up the digits and seeing whether they are equal to nine or a multiple of nine. Observe that 3 x 9 = 27 and the digits 2 + 7 = 9.

In this lesson, our focus is on learning all of the nine facts. Monitor the student's mastery, and proceed to the next lesson only when you are satisfied with his or her progress.

Example 1

$$9\overset{?}{\overset{\longrightarrow}{\big|63}} \qquad \frac{63}{9} = \qquad 63 \div 9 =$$

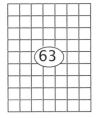

1. "What times nine is equal to 63?"
2. "Nine times what is equal to 63?"
3. "How many nines can I count out of 63?"
4. "63 divided by nine equals what number?"

Example 2

$$7\overset{?}{\overset{\longrightarrow}{\big|63}} \qquad \frac{63}{7} = \qquad 63 \div 7 =$$

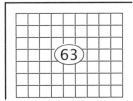

1. "What times seven is equal to 63?"
2. "Seven times what is equal to 63?"
3. "How many sevens can I count out of 63?"
4. "63 divided by seven equals what number?"

1 ÷ 1	2 ÷ 2	3 ÷ 3	4 ÷ 4	5 ÷ 5	6 ÷ 6	7 ÷ 7	8 ÷ 8	9 ÷ 9	10 ÷ 10
2 ÷ 1	4 ÷ 2	6 ÷ 3	8 ÷ 4	10 ÷ 5	12 ÷ 6	14 ÷ 7	16 ÷ 8	18 ÷ 9	20 ÷ 10
3 ÷ 1	6 ÷ 2	9 ÷ 3	12 ÷ 4	15 ÷ 5	18 ÷ 6	21 ÷ 7	24 ÷ 8	27 ÷ 9	30 ÷ 10
4 ÷ 1	8 ÷ 2	12 ÷ 3	16 ÷ 4	20 ÷ 5	24 ÷ 6	28 ÷ 7	32 ÷ 8	36 ÷ 9	40 ÷ 10
5 ÷ 1	10 ÷ 2	15 ÷ 3	20 ÷ 4	25 ÷ 5	30 ÷ 6	35 ÷ 7	40 ÷ 8	45 ÷ 9	50 ÷ 10
6 ÷ 1	12 ÷ 2	18 ÷ 3	24 ÷ 4	30 ÷ 5	36 ÷ 6	42 ÷ 7	48 ÷ 8	54 ÷ 9	60 ÷ 10
7 ÷ 1	14 ÷ 2	21 ÷ 3	28 ÷ 4	35 ÷ 5	42 ÷ 6	49 ÷ 7	56 ÷ 8	63 ÷ 9	70 ÷ 10
8 ÷ 1	16 ÷ 2	24 ÷ 3	32 ÷ 4	40 ÷ 5	48 ÷ 6	56 ÷ 7	64 ÷ 8	72 ÷ 9	80 ÷ 10
9 ÷ 1	18 ÷ 2	27 ÷ 3	36 ÷ 4	45 ÷ 5	54 ÷ 6	63 ÷ 7	72 ÷ 8	81 ÷ 9	90 ÷ 10
10 ÷ 1	20 ÷ 2	30 ÷ 3	40 ÷ 4	50 ÷ 5	60 ÷ 6	70 ÷ 7	80 ÷ 8	90 ÷ 9	100 ÷ 10

Finding the Area of a Parallelogram

To find the area of a rectangle, multiply the base times the height. This is the simplest formula, and we will use it, or a derivative of it, to find the area of other shapes in this book. The *height* is shown by the letter *h* and is always perpendicular to the *base*. This is the reason for the little square in the corner where the sides of the rectangle meet or intersect. Notice that the area of a shape is always found in square units or square feet, depending on what unit of measure you are using.

If you see one line to the upper right of a number, it indicates feet, as in 6' (6 feet). If you have two lines to the upper right of a number, then it indicates inches, as in 10" (10 inches). While the sides are measured in length, the area is measured in square units. When you use the blocks you can visualize the squares inside the rectangle as in example 1. Each square represents one square foot. I combine square and area to form "squarea" to remind me of this relationship.

Example 1

Find the area of the rectangle.

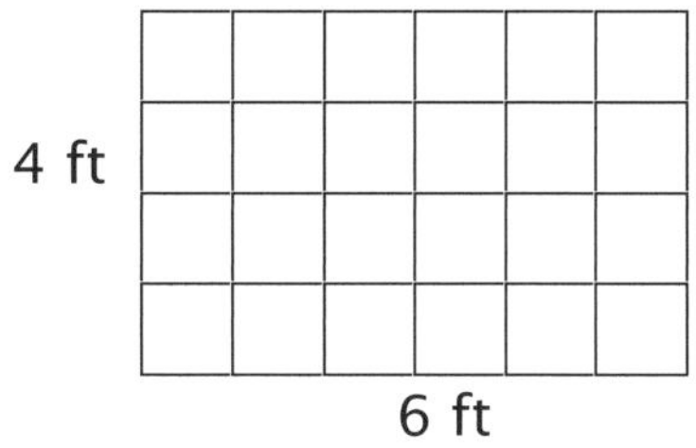

4 ft

6 ft

Area = bh = 6' x 4' = 24 square feet

The formula for finding the area of a *parallelogram* is the same as the formula for a rectangle, although you wouldn't think so when you first see the shape. Consider the drawing in figure 1 to help you see this. Notice how the triangle on the right is moved around and forms a rectangle. Notice also that the base and height remain the same in all the pictures.

Figure 1

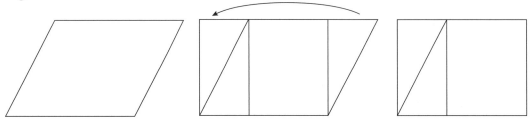

Example 2

Find the area of the parallelogram.

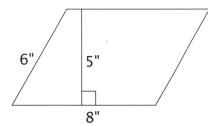

Area = bh = 8" x 5" = 40 square inches

Notice that the height in example 2 is 5" and not 6" because the height is always perpendicular to the base.

Example 3

Find the area of the parallelogram.

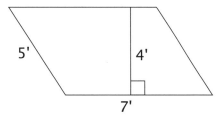

Area = bh = 7' x 4' = 28 square feet

The base can be measured on the top or on the bottom because they are the same length.

LESSON 8

Division by 6

Notice that all the multiples of six are even numbers. Notice also that when you add the digits of the multiples, they add up to three or a multiple of three. In 6 x 7 = 42, 42 is an even number and 4 + 2 = 6, which is a multiple of three. Carefully observe the student's progress and move to the next lesson only when you are satisfied with his or her mastery.

Example 1

$$\frac{24}{6} = \qquad 24 \div 6 =$$

1. "What times six is equal to 24?"
2. "Six times what is equal to 24?"
3. "How many sixes can I count out of 24?"
4. "24 divided by six equals what number?"

Example 2

$$\frac{24}{4} = \qquad 24 \div 4 =$$

1. "What times four is equal to 24?"
2. "Four times what is equal to 24?"
3. "How many fours can I count out of 24?"
4. "24 divided by four equals what number?"

1 ÷ 1	2 ÷ 2	3 ÷ 3	4 ÷ 4	5 ÷ 5	6 ÷ 6	7 ÷ 7	8 ÷ 8	9 ÷ 9	10 ÷ 10
2 ÷ 1	4 ÷ 2	6 ÷ 3	8 ÷ 4	10 ÷ 5	12 ÷ 6	14 ÷ 7	16 ÷ 8	18 ÷ 9	20 ÷ 10
3 ÷ 1	6 ÷ 2	9 ÷ 3	12 ÷ 4	15 ÷ 5	18 ÷ 6	21 ÷ 7	24 ÷ 8	27 ÷ 9	30 ÷ 10
4 ÷ 1	8 ÷ 2	12 ÷ 3	16 ÷ 4	20 ÷ 5	24 ÷ 6	28 ÷ 7	32 ÷ 8	36 ÷ 9	40 ÷ 10
5 ÷ 1	10 ÷ 2	15 ÷ 3	20 ÷ 4	25 ÷ 5	30 ÷ 6	35 ÷ 7	40 ÷ 8	45 ÷ 9	50 ÷ 10
6 ÷ 1	12 ÷ 2	18 ÷ 3	24 ÷ 4	30 ÷ 5	36 ÷ 6	42 ÷ 7	48 ÷ 8	54 ÷ 9	60 ÷ 10
7 ÷ 1	14 ÷ 2	21 ÷ 3	28 ÷ 4	35 ÷ 5	42 ÷ 6	49 ÷ 7	56 ÷ 8	63 ÷ 9	70 ÷ 10
8 ÷ 1	16 ÷ 2	24 ÷ 3	32 ÷ 4	40 ÷ 5	48 ÷ 6	56 ÷ 7	64 ÷ 8	72 ÷ 9	80 ÷ 10
9 ÷ 1	18 ÷ 2	27 ÷ 3	36 ÷ 4	45 ÷ 5	54 ÷ 6	63 ÷ 7	72 ÷ 8	81 ÷ 9	90 ÷ 10
10 ÷ 1	20 ÷ 2	30 ÷ 3	40 ÷ 4	50 ÷ 5	60 ÷ 6	70 ÷ 7	80 ÷ 8	90 ÷ 9	100 ÷ 10

Finding the Area of a Triangle

Remember the formula for finding the area of a rectangle or a parallelogram is base times height. The height is shown by the letter *h* and is always perpendicular to the base.

Example 1

Find the area of the rectangle.

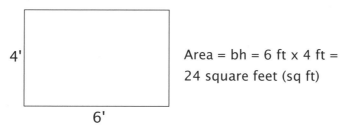

Area = bh = 6 ft x 4 ft =
24 square feet (sq ft)

Example 2

Find the area of the parallelogram.

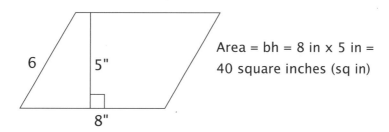

Area = bh = 8 in x 5 in =
40 square inches (sq in)

Notice that the height in example 2 is five inches and not six inches because the height is always perpendicular to the base.

A *triangle* is half of a rectangle or parallelogram. So the formula for the area of a triangle is one-half the area of the rectangle or parallelogram or one-half the base times the height, or 1/2·b·h. This can also be written as base times height divided by two, which is $\frac{b \times h}{2}$ or $\frac{bh}{2}$.

Taking one-half of a number is the same as dividing the number by two. Use whichever formula works best for you.

Example 3
Find the area of the unshaded triangle.

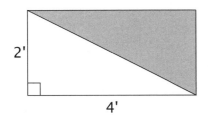

Area = 1/2 bh = 1/2 (4 x 2)=
1/2 (8) = 4 sq ft
or
Area = bh / 2 = (4 x 2) ÷ 2 =
(8 ÷ 2) = 4 sq ft

Example 4
Find the area of the unshaded triangle.

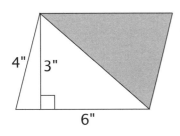

Area = 1/2 bh = 1/2 (6 x 3) =
1/2 (18) = 9 sq in
or
Area = bh / 2 = (6 x 3) ÷ 2 =
(18 ÷ 2) = 9 sq in

Example 5
Find the area of the triangle.

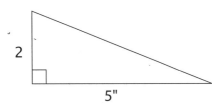

Area = 1/2 bh = 1/2 (5 x 2) =
1/2 (10) = 5 sq in
or
Area = bh / 2 = (5 x 2) ÷ 2 =
(10 ÷ 2) = 5 sq in

Example 6
Find the area of the triangle.

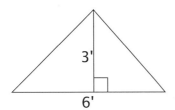

Area = 1/2 bh = 1/2 (6 x 3) =
1/2 (18) = 9 sq ft
or
Area = bh / 2 = (6 x 3) ÷ 2 =
(18 ÷ 2) = 9 sq ft

Division by 4

There aren't many tricks left for the four facts, but we don't need many since there are only three new facts that we haven't been exposed to as inverses of the facts we have already learned. If you learn these quickly, move on to the next lesson. In the problems with the box, have the student write the missing factor on the top, and then multiply and put the product under the number in the box and subtract as in figure 1. We're getting ready for more difficult problems.

Figure 1

$$
\begin{array}{r} ? \\ 4\,\overline{)\,24} \end{array}
\quad\rightarrow\quad
\begin{array}{r} 6 \\ .4\,\overline{)\,24} \end{array}
\quad\rightarrow\quad
\begin{array}{r} 6 \\ 4\,\overline{)\,24} \\ -24 \\ \hline 0 \end{array}
$$

Example 1

$$
\begin{array}{c} ? \\ 4\,\uparrow\,32 \end{array}
\qquad
\frac{32}{4} =
\qquad
32 \div 4 =
\qquad
4
$$

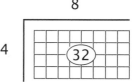

1. "What times four is equal to 32?"
2. "Four times what is equal to 32?"
3. "How many fours can I count out of 32?"
4. "32 divided by four equals what number?"

Example 2

$$4 \overset{?}{\uparrow} 28 \qquad \frac{28}{4} = \qquad 28 \div 4 = \qquad 4\,\overline{}^{\;7}$$

1. "What times four is equal to 28?"
2. "Four times what is equal to 28?"
3. "How many fours can I count out of 28?"
4. "28 divided by four equals what number?"

$1 \div 1$	$2 \div 2$	$3 \div 3$	$4 \div 4$	$5 \div 5$	$6 \div 6$	$7 \div 7$	$8 \div 8$	$9 \div 9$	$10 \div 10$
$2 \div 1$	$4 \div 2$	$6 \div 3$	$8 \div 4$	$10 \div 5$	$12 \div 6$	$14 \div 7$	$16 \div 8$	$18 \div 9$	$20 \div 10$
$3 \div 1$	$6 \div 2$	$9 \div 3$	$12 \div 4$	$15 \div 5$	$18 \div 6$	$21 \div 7$	$24 \div 8$	$27 \div 9$	$30 \div 10$
$4 \div 1$	$8 \div 2$	$12 \div 3$	$16 \div 4$	$20 \div 5$	$24 \div 6$	$28 \div 7$	$32 \div 8$	$36 \div 9$	$40 \div 10$
$5 \div 1$	$10 \div 2$	$15 \div 3$	$20 \div 4$	$25 \div 5$	$30 \div 6$	$35 \div 7$	$40 \div 8$	$45 \div 9$	$50 \div 10$
$6 \div 1$	$12 \div 2$	$18 \div 3$	$24 \div 4$	$30 \div 5$	$36 \div 6$	$42 \div 7$	$48 \div 8$	$54 \div 9$	$60 \div 10$
$7 \div 1$	$14 \div 2$	$21 \div 3$	$28 \div 4$	$35 \div 5$	$42 \div 6$	$49 \div 7$	$56 \div 8$	$63 \div 9$	$70 \div 10$
$8 \div 1$	$16 \div 2$	$24 \div 3$	$32 \div 4$	$40 \div 5$	$48 \div 6$	$56 \div 7$	$64 \div 8$	$72 \div 9$	$80 \div 10$
$9 \div 1$	$18 \div 2$	$27 \div 3$	$36 \div 4$	$45 \div 5$	$54 \div 6$	$63 \div 7$	$72 \div 8$	$81 \div 9$	$90 \div 10$
$10 \div 1$	$20 \div 2$	$30 \div 3$	$40 \div 4$	$50 \div 5$	$60 \div 6$	$70 \div 7$	$80 \div 8$	$90 \div 9$	$100 \div 10$

LESSON 11

Finding the Average

Isaiah 40:4 explains that every valley shall be exalted and every hill brought low. When I came across this, it reminded me of finding averages with the manipulatives. If I wanted to find the average age of three children who are 2, 3, and 10, I would get the blocks to represent each of the ages as in example 1.

Example 1

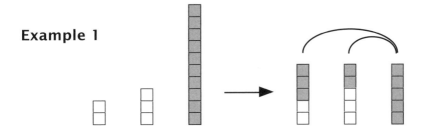

First I break up the 10 into smaller pieces using the one, two, and three unit bars. Then I move the blocks from the larger one onto the smaller ones until they are the same height. The valleys (two and three) are filled and the hill (10) is brought low.

The average age of the children is five. To do the same problem without blocks, add the ages of the children and divide by the number of children, as in example 2.

Example 2

2 + 3 + 10 = 15   15 ÷ 3 = 5

Example 3

During the past four months, we received the following amount of rainfall: 3 in, 7 in, 9 in, and 5 in. What is the average amount of rain per month?

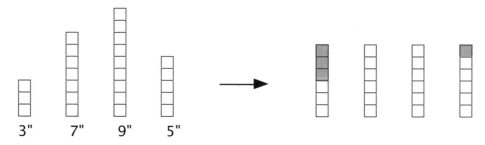

3 + 7 + 9 + 5 = 24 in

24 (total inches) ÷ 4 (number of months) = 6 inches per month

Division by 7 and 8

These are the last facts to be learned. There are four multiplication facts for which you have not learned the inverses. They are $7 \cdot 7 = 49$, $7 \times 8 = 56$, and $(8)(8) = 64$. When you learn the inverses of these, you will know all of the single-digit division facts. Then you will be ready for multiple-digit division.

Continue to write out the answers to problems in the box as in figure 1.

Figure 1

$$8 \overline{\smash{\big)}\ 56}^{\ ?} \quad \longrightarrow \quad 8 \overline{\smash{\big)}\ 56}^{\ 7} \quad \longrightarrow \quad \begin{array}{r} 7 \\ 8\overline{\smash{\big)}\ 5\ 6} \\ -5\ 6 \\ \hline 0 \end{array}$$

Example 1

$$7 \underset{49}{\overset{?}{\nearrow}} \qquad \frac{49}{7} = \qquad 49 \div 7 = \qquad 7 \boxed{49}^{\ 7}$$

Example 2

$$7 \underset{56}{\overset{?}{\nearrow}} \qquad \frac{56}{7} = \qquad 56 \div 7 = \qquad 7 \boxed{56}^{\ 8}$$

Example 3

$$8 \overset{?}{\underset{\uparrow}{|}} 64 \qquad \frac{64}{8} = \qquad 64 \div 8 = $$

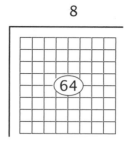

1 ÷ 1	2 ÷ 2	3 ÷ 3	4 ÷ 4	5 ÷ 5	6 ÷ 6	7 ÷ 7	8 ÷ 8	9 ÷ 9	10 ÷ 10
2 ÷ 1	4 ÷ 2	6 ÷ 3	8 ÷ 4	10 ÷ 5	12 ÷ 6	14 ÷ 7	16 ÷ 8	18 ÷ 9	20 ÷ 10
3 ÷ 1	6 ÷ 2	9 ÷ 3	12 ÷ 4	15 ÷ 5	18 ÷ 6	21 ÷ 7	24 ÷ 8	27 ÷ 9	30 ÷ 10
4 ÷ 1	8 ÷ 2	12 ÷ 3	16 ÷ 4	20 ÷ 5	24 ÷ 6	28 ÷ 7	32 ÷ 8	36 ÷ 9	40 ÷ 10
5 ÷ 1	10 ÷ 2	15 ÷ 3	20 ÷ 4	25 ÷ 5	30 ÷ 6	35 ÷ 7	40 ÷ 8	45 ÷ 9	50 ÷ 10
6 ÷ 1	12 ÷ 2	18 ÷ 3	24 ÷ 4	30 ÷ 5	36 ÷ 6	42 ÷ 7	48 ÷ 8	54 ÷ 9	60 ÷ 10
7 ÷ 1	14 ÷ 2	21 ÷ 3	28 ÷ 4	35 ÷ 5	42 ÷ 6	49 ÷ 7	56 ÷ 8	63 ÷ 9	70 ÷ 10
8 ÷ 1	16 ÷ 2	24 ÷ 3	32 ÷ 4	40 ÷ 5	48 ÷ 6	56 ÷ 7	64 ÷ 8	72 ÷ 9	80 ÷ 10
9 ÷ 1	18 ÷ 2	27 ÷ 3	36 ÷ 4	45 ÷ 5	54 ÷ 6	63 ÷ 7	72 ÷ 8	81 ÷ 9	90 ÷ 10
10 ÷ 1	20 ÷ 2	30 ÷ 3	40 ÷ 4	50 ÷ 5	60 ÷ 6	70 ÷ 7	80 ÷ 8	90 ÷ 9	100 ÷ 10

Mental Math

These problems can be used to keep the facts alive in the memory and to develop mental-math skills. Say the problem slowly enough so that the student comprehends, and then walk him through increasingly difficult exercises. The purpose is to stretch but not discourage. You decide where that line is!

Example 4

"Two plus three times four equals what number?"

The student thinks, "Two plus three equals five, and five times four equals twenty." At first, you will need to go slowly enough for him or her to verbalize the intermediate step. As skills increase, the student should be able to give just the answer.

The questions are intended to be read aloud in the order written. Do not worry about order of operations. These questions include addition, subtraction, and multiplication. Division will be included in lesson 18.

You will find more suggested mental-math problems for you to read aloud to your student in lessons 18 and 24. Try a few at a time, and remember to go quite slowly at first. Be sure your student is comfortable with the shorter problems before trying the longer ones.

1. Two plus six minus five equals what number? (3)

2. Seven minus four times nine equals what number? (27)

3. Two times four minus six equals what number? (2)

4. Seven plus six minus four equals what number? (9)

5. Four plus three times seven equals what number? (49)

6. Two times three minus one plus eight equals what number? (13)

7. Five plus seven minus six times seven equals what number? (42)

8. Seventeen minus nine plus one times six equals what number? (54)

9. Seven plus three times four plus three equals what number? (43)

10. Three times three minus seven times eight equals what number? (16)

Finding the Area of a Trapezoid

A *trapezoid* is a *quadrilateral* (four-sided figure) with two sides parallel. In the picture, notice that the top and bottom are parallel, but the sides are not. The top and bottom are called the *bases*. Finding the area of a trapezoid is a derivative of finding the area of a rectangle. The area of a rectangle is found by multiplying the base times the height. The formula for the area of a trapezoid is the *average* base times the height. Consider figure 1 to see where this formula originates.

Figure 1

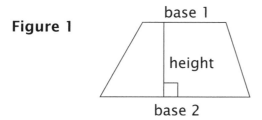

On the left and right sides, we choose a point in the middle and make one small triangle on each side. Then we pivot on this midpoint and swing the lower triangles up on both sides to make a rectangle out of the trapezoid. The resulting base is the average of the top and bottom bases and is found by connecting the two midpoints on both sides.

The traditional formula for finding the area of a trapezoid is $\frac{b_1+b_2}{2} \times h$.

The average base is found by adding the top and bottom bases and dividing by two.

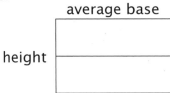

Then this result is multiplied by the height to find the area.

Example 1

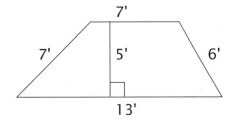

Find the area of the trapezoid.
The average base is
$\qquad$ (7 + 13) ÷ 2 = 10 ft.
The height is 5 ft.
The area is 10 x 5 =
$\qquad$ 50 square feet.

Example 2

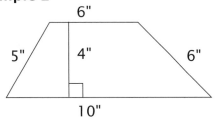

Find the area of the trapezoid.
The average base is
$\qquad$ (6 + 10) ÷ 2 = 8 in.
The height is 4 in.
The area is 8 x 4 =
$\qquad$ 32 square inches.

Thousands, Millions, and Place-Value Notation

A huge component in understanding multiple-digit division is place value. We call the beginning value units. This is represented by the small, green one-half-inch cube. The next largest place value is the tens place, shown with the blue 10 bar. This is 10 times as large as a unit. The next value we come to, as we move to the left, is the hundreds place, the large red block. It is 10 times as large as the tens. Notice that as you move to the left, each value is 10 times as large as the preceding value. See figure 1.

Figure 1

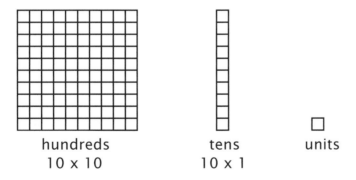

hundreds
10 x 10

tens
10 x 1

units

When you name a number such as 247, the 2 tells you how many hundreds, the 4 indicates how many tens, and the 7, how many units. We read 247 as "Two hundred for-ty (ty means ten) seven." The 2, 4, and 7 are digits that tell us *how many*. The hundreds, tens, and units tell us what kind, or *what value*. Where the digit is written, or what *place* it occupies, tells us what *value* it is.

Notice that as the values progress from right to left, they increase by a factor of 10. That is because we are operating in the base 10 system, otherwise known as the decimal system.

The next place value after the hundreds is the thousands place. It is 10 times 100. You could build a 1,000 by stacking 10 hundred squares and making a cube. You can also show a 1,000 by making a rectangle that is 10 by 100 out of the cube as in figure 2. You will see that I made the picture to a much smaller scale to be able to show this. Figure 2 also shows 10,000. Can you imagine what 100,000 would look like if you stick to rectangles? It would be a rectangle 100 by 1,000. The factors are noted inside the rectangles in the drawing.

Figure 2

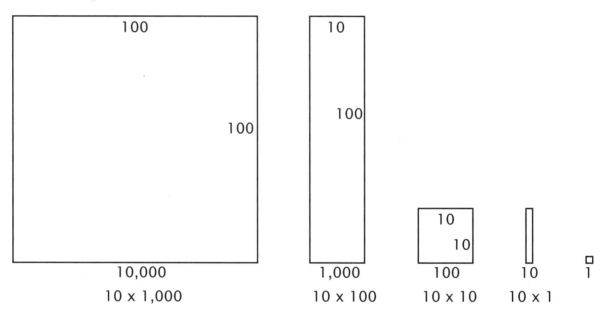

There is a lot of information in figure 2. Notice the progression of multiplying by a factor of 10 as you move from right to left. Ten times 1,000 is 10,000, and 10 times 10,000 is 100,000. Finally, 10 times 100,000 is 1,000,000 (one million). Do you also see that within the thousands group there are units, tens, and hundreds places, just as in the units group (figure 1)? The same is true with the millions. There are one million, ten million, and one hundred million places. The commas separate the units, thousands, and millions. See figure 3.

Figure 3

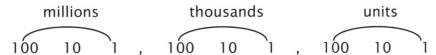

When saying these larger numbers, I like to think of the commas as having names. The first comma from the right is named "thousand" and the second from the right is "million." See example 1.

Example 1
Say 123,456,789.

"123 million 456 thousand 789"   or
"one hundred twenty-three million, four hundred fifty-six thousand, seven hundred eighty-nine"

Notice that you never say "and" when reading a large number. This is reserved for the decimal point in a succeeding book. Do you see that you never read a number larger than hundreds between the commas? This is because there are only three places between the commas. Practice saying and writing larger numbers.

Example 2
Say 13,762.

"13 thousand 762"   or
"thirteen thousand, seven hundred sixty-two"

Place-value notation is a way of writing numbers that emphasizes the place value. The number in example 2, which is 13,762, looks like this in place-value notation: 10,000 + 3,000 + 700 + 60 + 2. Each value is written separately.

Example 3
Write 8,543,971 with place-value notation.

8,000,000 + 500,000 + 40,000 + 3,000 + 900 + 70 + 1

Billions, Trillions, and Expanded Notation
Multi-Step Word Problems

The next two sections beyond the millions are the billions and trillions. See figure 1 and example 1.

Figure 1

trillions	billions	millions	thousands	units
100 10 1 ,	100 10 1 ,	100 10 1 ,	100 10 1 ,	100 10 1

Example 1

Say 135,246,123,456,789.

"135 trillion 246 billion 123 million 456 thousand 789" or "one hundred thirty-five trillion, two hundred forty-six billion, one hundred twenty-three million, four hundred fifty-six thousand, seven hundred eighty-nine."

We have already learned that as the values progress from right to left, they increase by a factor of 10. There is another pattern that occurs as you move from right to left between the place values. See figure 2.

Take the first number to the left of each comma as an example. Begin with one unit, then one thousand, one million, one billion, and one trillion. Because there are three spaces between each arrow, this should be 10 x 10 x 10, or 1,000, times as large as the preceding one. The first arrow points at 1, the next arrow to the left is 1,000, the next arrow is pointing at 1,000,000, etc.

Figure 2

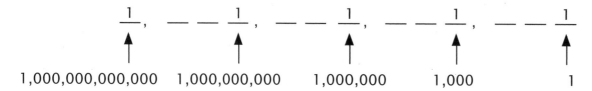

A thousand is a thousand units.
A million is a thousand thousands.
A billion is a thousand millions.
A trillion is a thousand billions.

Expanded Notation - Another notation to learn is ***expanded notation***. It separates the digit and the place value and goes one step further than place-value notation. See example 2.

Example 2
Write 43,971 in expanded notation form.

4x10,000 + 3x1,000 + 9x100 + 7x10 + 1

Multi-Step Word Problems

The student text includes some fairly simple two-step word problems. Some students may be ready for more challenging problems. Here are a few to try, along with some tips for solving this kind of problem. You may want to read and discuss these with your student as you work out the solutions together. Again, the purpose is to stretch, not to frustrate. If you do not think the student is ready, you may want to come back to these later.

There are more multi-step word problems in lessons 21 and 27 of the instruction manual. The answers are at the end of the solutions at the back of this book.

1. David has a rectangular garden that measures 11 feet by 13 feet. He wants to plant peas in his garden. Dad said that one seed packet will be enough to fill a space 10 feet on a side. Will David's garden have enough space to plant two seed packets?

Although the problem asks only one question, there are other questions that must be answered first. The key to solving this is determining what the unstated questions are. Since the final question is really asking for a comparison of the available area to the needed area, the two unstated questions are: "What is the area of David's garden?" and "What is the area needed for two seed packets?"

You might make a list of steps something like this:

1. area of garden in square feet?
2. area needed for one seed packet?
3. area needed for two seed packets?
4. compare areas to answer question

2. Rachel and Sarah started out to visit Grandma. They drove for 50 miles and stopped to rest before driving for 30 more miles. They decided to go back 10 miles to a restaurant they had seen. After leaving the restaurant, they drove 80 more miles to Grandma's house. How many miles did the girls drive on the way to Grandma's house?

Make a drawing, and this will be easier to solve!

3. Rachel and Sarah spent $8 for gasoline, $15.65 for their lunch, and $5 apiece for gifts for Grandma. Grandma gave each of them $10. If the girls left home with a total of $50, how much do they have for the return trip?

This is similar to number 1 in that you must answer other questions before you can answer the question in the problem.

Division by a Single Digit with Remainder

While we have been dividing for some time, all of the answers have come out even. In real life, not all problems come out even. For example, what if I have 13 quarters? How many dollars is that? The problem is, "How many dollars, or groups of four quarters, are there in 13 quarters?" Or, "13 ÷ 4 is equal to what number?" Look at example 1 to see how we work out with numbers and symbols and with the blocks when there is a *remainder*.

Example 1

$$\begin{array}{r} 3 \\ 4\overline{)13} \end{array}$$

How many groups of 4 will go into 13? The answer is 3.

$$\begin{array}{r} 3 \\ 4\overline{)13} \\ -12 \\ \hline 1 \end{array}$$

Multiply. 3 x 4 =12. Place the 12 under the 13 and subtract it.

$$\begin{array}{r} 3 \text{ r. } 1 \\ 4\overline{)13} \\ -12 \\ \hline 1 \end{array}$$

13 minus 12 leaves 1, which is our remainder. We write this as "r.1" for remainder 1.

Notice in example 1 that the remainder can not be larger than 3. If the remainder was 4, then you would have another group of 4, and 16 divided by 4 would be 4. The blocks reveal that the remainder must be a 0, 1, 2, or 3. All the problems that have a whole number answer, or come out even, have a remainder of zero. Consider the next few examples.

Example 2

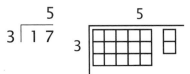

How many groups of 3 will go into 17? The answer is 5.

$$\begin{array}{r} 5 \\ 3\overline{)17} \\ -15 \\ \hline 2 \end{array}$$

Multiply. 5 x 3 = 15. Place the 15 under the 17 and subtract it.

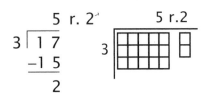

17 minus 15 leaves 2, which is our remainder. We write this as "r.2" for remainder 2.

Example 3

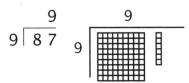

How many groups of 9 will go into 87? The answer is 9.

$$\begin{array}{r} 9 \\ 9\overline{)87} \\ -81 \\ \hline 6 \end{array}$$

Multiply. 9 x 9 = 81. Place the 8 under the 87 and subtract it.

$$\begin{array}{r} 9 \text{ r. } 6 \\ 9\overline{)87} \\ -81 \\ \hline 6 \end{array}$$

87 minus 81 leaves 6, which is our remainder. We write this as "r.6" for remainder 6.

LESSON 17

Upside-Down Multiplication

The student should already know how to multiply multiple-digit numbers. This lesson tells how to do it upside down in order to review the role of place value in multiplication, and to prepare us for division. Even though it will take up more space on your paper, it is easier to keep track of the place value. As you study the following examples, take special note of the place value. We'll do each of the problems with regular notation as well as place-value notation. When possible, we'll use the blocks simultaneously. Example 1 is the traditional way of multiplying multiple-digit numbers and example 2 is the upside-down way.

Example 1

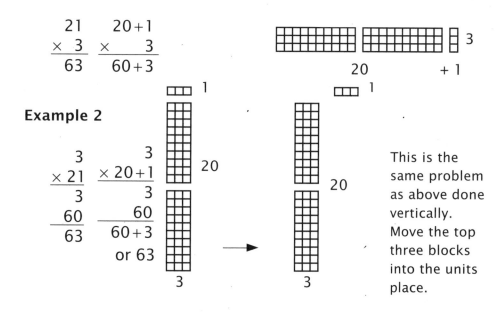

Example 2

This is the same problem as above done vertically. Move the top three blocks into the units place.

Example 3

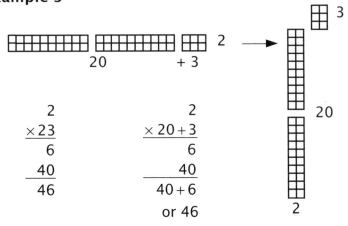

```
      2              2
    ×23          ×20+3
      6              6
     40             40
     46           40+6

                  or 46
```

Example 4

```
      3              3
   ×132       ×100+30+2
      6              6
     90             90
    300            300
    396            396

              or 300 + 90 + 6
```

Example 4 (traditional)

```
    132        100+30+2
  ×   3        ×       3
    396        100+90+6
```

Example 5

```
      6              6
  × 218       ×200+10+8
     48             48
     60             60
   1200           1200
   1308           1308
```

Example 5 (traditional)

$$
\begin{array}{r}
218 \\
\times\ \ \ 6 \\
\hline
4 \\
1264 \\
\hline
1308
\end{array}
\qquad
\begin{array}{r}
200+10+8 \\
\times \qquad\qquad 6 \\
\hline
40 \\
1000+200+60+8 \\
\hline
1000+300+00+8
\end{array}
$$

Another stepping stone to understanding multiple-digit division is shown in examples 6–10. The student knows how to multiply numbers with tens and hundreds and knows his division facts. This step puts the two together and prepares the student for more difficult problems.

Example 6

$$
\begin{array}{r}
30 \\
3\overline{\smash{)}90} \\
-90 \\
\hline
0
\end{array}
$$

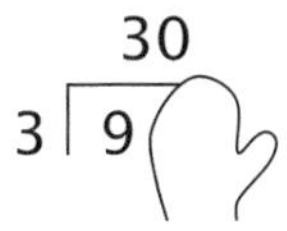

Example 7

$$
\begin{array}{r}
100 \\
7\overline{\smash{)}700} \\
-700 \\
\hline
0
\end{array}
$$

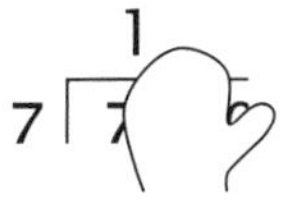

Example 8

$$
\begin{array}{r}
200 \\
4\overline{\smash{)}800} \\
-800 \\
\hline
0
\end{array}
$$

Example 9

$$
\begin{array}{r}
40 \\
6\overline{\smash{)}240} \\
-240 \\
\hline
0
\end{array}
$$

Example 10

$$
\begin{array}{r}
70 \\
9\overline{\smash{)}630} \\
-630 \\
\hline
0
\end{array}
$$

Division with Double-Digit Factors

In each of the examples, we will walk the student through each step. Notice that when dividing, you begin by looking for the largest amount to multiply and subtract, and then you proceed to the smaller amounts.

When multiplying, we do the opposite and multiply the units first, then the tens, then the hundreds, and so forth.

This is the reverse order, and it should not be surprising because division is the inverse of multiplication. Notice how we work each problem with the manipulatives, and then with place-value notation and regular notation.

Example 1

$$
\begin{array}{r}
20 \\
2\,\overline{)\,40} \\
-40 \\
\hline
0
\end{array}
\qquad\qquad
\begin{array}{r}
3 \\
2\,\overline{)\,6} \\
-6 \\
\hline
0
\end{array}
$$

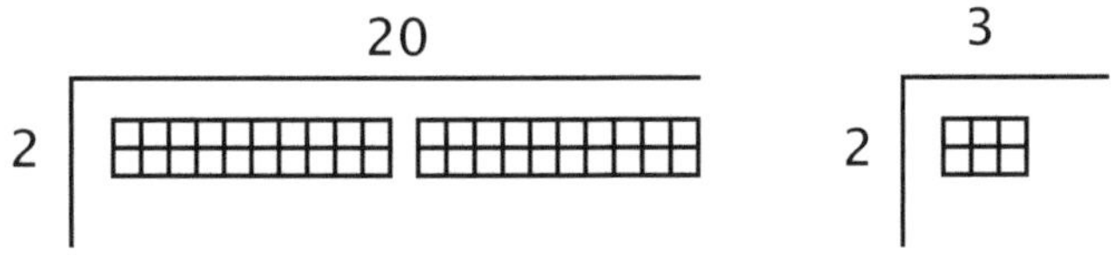

Example 1 shows two small problems, 40 ÷ 2 and 6 ÷ 2. In example 2, the smaller problems are combined. First we divide the tens place, and then the units place. The two-step process is new.

Example 2

```
        3
       20        20 + 3
    2 | 46     2 | 40 + 6
      −40          −40
        6            6
       −6           −6
        0            0
```

How many groups of 2 will go into 40? The answer is 20.

Multiply: 2 x 20 = 40. Place 40 under the 46 and subtract it. 46 minus 40 leaves 6.

How many groups of 2 will go into 6? The answer is 3.

Multiply: 2 x 3 = 6. Place 6 under the 6 and subtract it. There is no remainder.

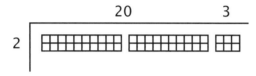

Always check your answer by multiplying.

```
           2                2
    ×   20 + 3        ×    23
           6                6
          40               40
       40 + 6  or 46       46
```

Example 3

```
        2
       10        10 + 2
    4 | 48     4 | 40 + 8
      −40          −40
        8            8
       −8           −8
        0            0
```

How many groups of 4 will go into 40? The answer is 10.

Multiply: 4 x 10 = 40. Place 40 under the 48 and subtract it. 48 minus 40 leaves 8.

How many groups of 4 will go into 8? The answer is 2.

Multiply: 4 x 2 = 8. Place 8 under the 8 and subtract it. There is no remainder.

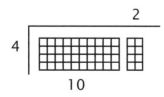

Always check your answer by multiplying.

```
           4                4
    ×   10 + 2        ×    12
           8                8
          40               40
       40 + 8  or 48       48
```

Example 4

```
           2
          30              30+2
       3 ⟌98           3 ⟌90+8
         −90             −90
           8               8
          −6              −6
           2               2
```

How many groups of 3 will go into 90? The answer is 30.

Multiply: 3 x 30 = 90. Place 90 under the 98 and subtract it. 98 minus 90 leaves 8.

How many groups of 3 will go into 8? The answer is 2.

Multiply: 3 x 2 = 6. Place 6 under the 8 and subtract it. There is a remainder of 2.

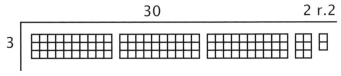

Always check your answer by multiplying.

```
                    3              3
         ×       30+2        ×   32
                    6              6
                   90             90
         90+6+2=98             96
                                   2
                                  98
```

Example 5

```
           8
          20              20+8
       2 ⟌56           2 ⟌50+6
         −40             −40
          16             10+6
         −16            −10+6
           0               0
```

How many groups of 2 will go into 50? The answer is 20.

Multiply: 2 x 20 = 40. Place 40 under the 56 and subtract it. 56 minus 40 leaves 16.

How many groups of 2 will go into 16? The answer is 8.

Multiply: 2 x 8 = 16. Place 16 under the 16 and subtract it. There is no remainder.

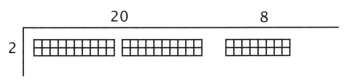

Always check your answer by multiplying.

```
                    2              2
         ×       20+8        ×  28
                   16             16
                   40             40
         40+16=56              56
```

Example 6

$$
\begin{array}{r}
9 \\
10 \\
5\,\overline{)\,95} \\
-50 \\
\hline
45 \\
-45 \\
\hline
0
\end{array}
\qquad
\begin{array}{r}
10+9 \\
5\,\overline{)\,90+5} \\
-50 \\
\hline
40+5 \\
-40+5 \\
\hline
0
\end{array}
$$

How many groups of 5 will go into 90?
The answer is 10.
  Multiply: 5 x 10 = 50. Place 50 under the 95 and subtract it. 95 minus 50 leaves 45.

How many groups of 5 will go into 45?
The answer is 9.
  Multiply: 5 x 9 = 45. Place 45 under the 45 and subtract it. There is no remainder.

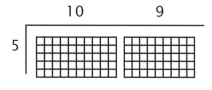

Always check your answer by multiplying.

$$
\begin{array}{r}
5 \\
\times 10+9 \\
\hline
45 \\
50 \\
\hline
95
\end{array}
\qquad
\begin{array}{r}
5 \\
\times 19 \\
\hline
45 \\
50 \\
\hline
95
\end{array}
$$

Mental Math

Here are some more mental math problems for you to read aloud to your student. Try a few at a time, going slowly at first. Be sure your student is comfortable with the shorter problems before trying the longer ones.

1. Forty-two divided by six plus five equals ? (12)

2. Three times four divided by six equals ? (2)

3. Thirteen minus five divided by two equals ? (3)

4. Seven plus eight divided by three equals ? (5)

5. Twenty minus ten divided by five equals ? (2)

6. Six plus six minus three divided by three equals ? (3)

7. Forty-five divided by five plus one plus ten equals ? (20)

8. Eighteen minus nine plus seven divided by two equals ? (8)

9. Two times three times four divided by eight equals ? (3)

10. Seventy-two divided by eight divided by three times seven equals ? (21)

LESSON 19

Division, Three Digit by One Digit

The first question we ask when dividing is, "How many groups?" Even though that is correct, it could be restated. In example 1, we are looking for how many groups of 3 will go into 900. Since this is the hundreds place, we could look for the multiple of 3 times 100, or 300, that goes into 900. When dividing into the hundreds place, this is the place to begin, with 100 times the outside factor (3 in this case). As we work through the problem, the next place value is the tens. We should be looking for what times 30 (3 x 10) will go into 36. Then when we come to the units, we have the easy one: what times 3 (3 x 1) goes into 6.

Example 1

```
          2
         10
        300
     3 ) 936
        -900
         36
        -30
          6
         -6
```

What is the hundreds multiple times 3 that goes into 900? It is 300. Multiply: 3 x 300 = 900. Place 900 under the 936 and subtract it. 936 minus 900 leaves 36.

What is the tens multiple times 3 that goes into 30? It is 10. Multiply: 3 x 10 = 30. Place 30 under the 36 and subtract it. 36 minus 30 leaves 6.

Now what is the units multiple times 3 that goes into 6? It is 2. Multiply: 3 x 2 = 6. Place 6 under the 6 and subtract it. There is no remainder.

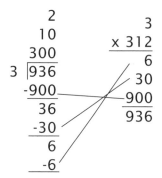

When you check your answer with multiplication, notice how the partial products of the multiplication problem correspond to the division progression.

Example 2

```
        8
       20
      200 r. 1
    2 |457
     -400
       57
      -40
       17
      -16
        1
```

What is the hundreds multiple times 2 that goes into 400? It is 200. Multiply: 2 x 200 = 400. Place 400 under the 457 and subtract it. 457 minus 400 is 57.

There is no tens multiple times 2 that makes 50. What is the tens multiple times 2 that goes into 40? It is 20. Multiply: 2 x 20 = 40. Place 40 under the 57 and subtract it. 57 minus 40 leaves 17.
Now what is the units multiple times 2 that goes into 17? It is 8. Multiply: 2 x 8 = 16. Place 16 under the 17 and subtract it. The remainder is 1.

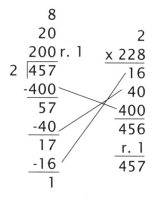

When you check your answer with multiplication, notice how the partial products of the multiplication problem correspond to the division progression.

In examples 3 and 4, we encounter two types of problems that can be difficult if you are just memorizing a formula. However, if we understand what we are doing, they should be clear. Read them through carefully.

Example 3

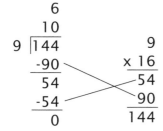

What is the hundreds multiple times 9 that goes into 100? It is 0. So we move to the tens place.

What is the tens multiple times 9 that goes into 140? It is 10. Multiply: 9 x 10 = 90. Place 90 under the 144 and subtract it. 144 minus 90 leaves 54.

Now what is the units multiple times 9 that goes into 54? It is 6. Multiply: 9 x 6 = 54. Place 54 under the 54 and subtract it. There is no remainder.

```
      6
     10
  9 |144        9
    -90      x 16
     54        54
    -54        90
      0       144
```

When you check your answer with multiplication, notice how the partial products of the multiplication problem correspond to the division progression.

Example 4

```
      7
      0
    100 r.2
  3 |323
    -300
     23
     -0
     23
    -21
      2
```

What is the hundreds multiple times 3 that goes into 300? It is 100. Multiply: 3 x 100 = 300. Place 300 under the 323 and subtract it. 323 minus 300 is 23.

What is the tens multiple times 3 that goes into 23? It is 0. There is no answer other than zero, so we put a zero in the tens place and proceed to the units.

Now what is the units multiple times 3 that goes into 23? It is 7. Multiply: 3 x 7 = 21. Place 21 under the 23 and subtract it. The remainder is 2.

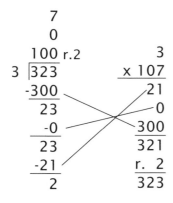

When you check your answer with multiplication, notice how the partial products of the multiplication problem correspond to the division progression.

LESSON 20

Division, Three Digit by One Digit
with Fraction Remainder

So far in our study of division, we have written remainders with "r." Another way to show what is left over is to show division as a fraction. In example 1, there is a simple division problem written with the house or rectangle and with a line between the two numbers to be divided.

Example 1

$$\begin{array}{r} 2 \text{ r. } 1 \\ 4\overline{)\,9} \\ \underline{-8} \\ 1 \end{array} \qquad \frac{9}{4} = \frac{8}{4} + \frac{1}{4} = 2 + \frac{1}{4} = 2\frac{1}{4} \qquad \begin{array}{r} 2\frac{1}{4} \\ 4\overline{)\,9} \\ \underline{-8} \\ 1 \end{array}$$

Verbalizing this problem helps to understand the new symbolism. Nine divided by four is two, with one left over. Then one divided by four or 1 ÷ 4 may be written as 1/4. So another way to write 2 r.1 is 2 1/4. Here is another example.

Example 2

$$\begin{array}{r} 4 \text{ r. } 5 \\ 7\overline{)\,33} \\ \underline{-28} \\ 5 \end{array} \qquad \frac{33}{7} = \frac{28}{7} + \frac{5}{7} = 4 + \frac{5}{7} = 4\frac{5}{7} \qquad \begin{array}{r} 4\frac{5}{7} \\ 7\overline{)\,33} \\ \underline{-28} \\ 5 \end{array}$$

For those who are concerned that we haven't studied *fractions* yet, do not be disconcerted. As you have seen in the first two examples, writing the remainder as "a line between the two numbers to be divided" is division. But the top number, or product, is smaller than the bottom number, or factor, and we can't divide any further, so we leave it in that form. It looks like a fraction, but it is simply a division problem gone as far as it can.

Example 3

```
        6
       70
      100  1
           ─
   2 │353  2
    -200
     153
    -140
      13
     -12
       1
```

What is the largest hundreds multiple times 2 that goes into 300? It is 200. Multiply: 2 x 100 = 200. Place 200 under the 353 and subtract it. 353 minus 200 leaves 153.

What is the largest tens multiple times 2 that goes into 153? It is 70. Multiply: 2 x 70 = 140. Place 140 under the 153 and subtract it. 153 minus 140 is 13.

Now what is the largest units multiple times 2 that goes into 13? It is 6. Multiply: 2 x 6 = 12. Place 12 under the 13 and subtract it. The remainder is 1.
We can also write the remainder as 1/2.

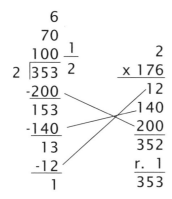

By this point, hopefully, the student understands the role of place value in division. Instead of writing the top factor separated into different place values, we can begin writing the number on one line. Don't forget the place value, and think of it as you solve for the missing factor. If the student is more comfortable with writing it all out as before, you, the teacher, should determine what is in his or her best interest. I still like to check by multiplying "upside down," but either way will work as shown in example 4.

Example 4

$$
\begin{array}{r}
132\ \tfrac{3}{6} \\
6\ \overline{)795}\ ^6 \\
-600 \\
\hline
195 \\
-180 \\
\hline
15 \\
-12 \\
\hline
3
\end{array}
$$

What is the largest hundreds multiple times 6 that goes into 795? It is 100. Multiply: 6 x 100 = 600. Place 600 under the 795 and subtract it. 795 minus 600 leaves 195.

What is the largest tens multiple times 6 that goes into 195? It is 30. Multiply: 6 x 30 = 180. Place 180 under the 195 and subtract it. 195 minus 180 is 15.

Now what is the largest units multiple times 6 that goes into 15? It is 2. Multiply: 6 x 2 = 12. Place 12 under the 15 and subtract it. The remainder is 3. We can also write the remainder as 3/6.

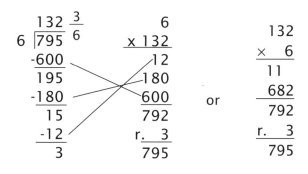

Example 5

$$167 \frac{4}{5}$$
$$5 \overline{|839|}^5$$
$$-500$$
$$\overline{339}$$
$$-300$$
$$\overline{39}$$
$$-35$$
$$\overline{4}$$

What is the largest hundreds multiple times 5 that goes into 839? It is 100. Multiply: 5 x 100 = 500. Place 500 under the 839 and subtract it. 839 minus 500 leaves 339.

What is the largest tens multiple times 5 that goes into 339? It is 60. Multiply: 5 x 60 = 300. Place 300 under the 339 and subtract it. 339 minus 300 is 39.

Now what is the largest units multiple times 5 that goes into 39? It is 7. Multiply: 5 x 7 = 35. Place 35 under the 39 and subtract it. The remainder is 4. We can also write the remainder as 4/5.

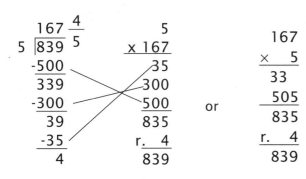

$$\begin{array}{r} 167 \\ \times\ 5 \\ \hline 33 \\ 505 \\ \hline 835 \\ \text{r.}\quad 4 \\ \hline 839 \end{array}$$

Rounding to 10, 100, 1,000, and Estimation

Most of this lesson should be review, as we have covered this material in the earlier books. If it is new, take your time and thoroughly digest it. Rounding to 10 is used in estimating when we multiply or divide. When you round a number to the nearest multiple of 10, there will be a number in the tens place, but only a zero in the units place. I tell the students this is why we call it rounding, because the units are going to be a "round zero."

Let's round 38 as an example. The first skill is to find the two multiples of 10 that are nearest to 38. The next-lower multiple is 30, and the next-higher multiple is 40. We see that 38 is between 30 and 40. If the student has trouble finding these numbers, begin by placing your finger over the eight in the units place, so that all you have is a three in the tens place. A three in the tens place is 30. Then add one more ten to find the 40. I often write the numbers 30 and 40 above the number 38 on both sides as in figure 1.

Figure 1   30   40
                        38

The next skill is find out whether 38 is closer to 30 or 40. Let's go through all the numbers as in figure 2. It is obvious that 31, 32, 33, and 34 are closer to 30, and 36, 37, 38, and 39 are closer to 40. But 35 is right in the middle. Somebody decided that it rounds to 40, so that is the reason for our rule. When rounding to tens, look at the units place. If the units are 0, 1, 2, 3, or 4, the digit in the tens place remains unchanged. If the units are 5, 6, 7, 8, or 9, the digit in the tens place increases by one. See figure 2.

Figure 2

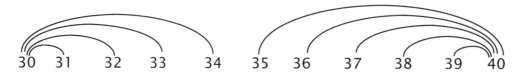

Another strategy I use is to put 0, 1, 2, 3, and 4 inside a circle to represent "0," because if these numbers are in the units place, they add nothing to the tens place, and they are rounded to the lower number (30 in the example). Then I put 5, 6, 7, 8, and 9 inside a thin rectangle to represent "1," because if these numbers are in the units place, they add one to the tens place and are rounded to the higher number (40 in the example).

Figure 3

Example 1

Round 43 to the nearest tens place.

40 50 43	1. Find the multiples of 10 nearest to 43.
40 ↞ 50 43	2. We know that three goes to the lower number, which is 40.
(40) ↞ 50 43	3. Or recall that three is in the circle (zero) so nothing is added to the smaller number, which is 40.

When rounding to hundreds, look only at the digit in the tens place to determine whether to stay the same or increase by one. The same rules apply to hundreds as to tens; if it is a 0, 1, 2, 3, or 4, the number in the hundreds place remains unchanged. If the digit in the tens place is a 5, 6, 7, 8, or 9, then the digit in the hundreds place increases by one. See example 2.

Example 2

Round 547 to the nearest hundreds place.

500 547 600

1. Find the multiples of 100 nearest to 547.

500 ← 600
  547

2. We know that four goes to the lower number, which is 500.

(500) ← 600
  547

3. Or recall that four is in the circle (zero) so nothing is added to the smaller number, which is 500.

When rounding to thousands, consider only the number immediately to the right of the thousands place, which is the hundreds, to determine whether to stay the same or increase by one. See example 3.

Example 3

Round 8,719 to the nearest thousands place.

8,000 8,719 9,000

1. Find the nearest multiples of 1,000.

8,000 → 9,000
  8,719

2. We know that seven goes to the higher number, which is 9,000.

8,000 → (9,000)
  8,719

3. Or recall that seven is in the rectangle (one) so one is added to the eight.

The answer is 9,000.

Estimation - Now that we know how to round numbers, we can apply this to find the approximate answer for a division problem. In example 4, we are just going to round the product inside the house. Later when the outside factor has multiple digits, we will round both of them. Round 627 to 600 first, and then use your hand to cover the zeros and quickly divide three into six to get two. Then add the zeros again to get 200. Finally, do the division and compare your approximation with the exact answer. The symbol ≈ means "approximately equal to."

Note: Getting the correct first number is not nearly as important as getting the correct place to put the first number. In other words, finding the estimated place value is the most important part of this exercise. In example 5, the first number will not turn out to be a two in the final answer, but as long as you know it will be in the hundreds place, that is what is most critical.

Example 4
Estimate the answer.

$$3\overline{)627} \rightarrow 3\overline{)(600)} \rightarrow 3\overline{)(6} \rightarrow 3\overline{)(600)}^{200}$$

Example 5
Estimate the answer.

$$4\overline{)781} \rightarrow 4\overline{)(800)} \rightarrow 4\overline{)(8} \rightarrow 4\overline{)(800)}^{200}$$

Example 6
Estimate the answer.

$$2\overline{)895} \rightarrow 2\overline{)(900)} \rightarrow 2\overline{)(9} \rightarrow 2\overline{)(900)}^{400}$$

Multi-Step Word Problems

Here are a few more multi-step word problems to try. You may want to read and discuss these with your student as you work out the solutions together. Again, the purpose is to stretch, not to frustrate. If you do not think the student is ready, you may want to come back to these later.

The answers are at the end of the solutions at the back of this book.

1.   Scott bought 3 bags of candy with 75 pieces in each one. He plans to divide all the candy evenly among seven friends. How many pieces of candy will Scott have left for himself?

2.   Anne earned $3 an hour babysitting and $4 an hour working in the garden. Last week she did babysitting for five hours and garden work for three hours. How much more money does she need to buy a game that costs $35?

3.   Riley had a nature collection. She had 25 acorns, 16 dried seed pods, and 8 feathers. She divided the acorns into five equal groups, the seed pods into four equal groups, and the feathers into two equal groups. She gave her mother one group of each kind. How many separate items did her mother get?

Division, Three Digit by Two Digit

The opposite of multiplying 12 x 13 = 156 is dividing: 156 ÷ 12 = 13 or 156 ÷ 13 = 12. We'll do a couple of examples with the blocks, as well as with the written notation, to review this relationship.

Example 1
Solve 156 ÷ 13.

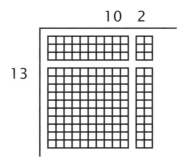

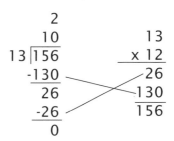

Example 2
Solve 276 ÷12.

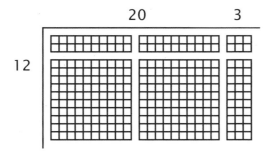

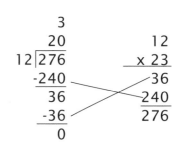

Example 3

Solve 387 ÷ 11.

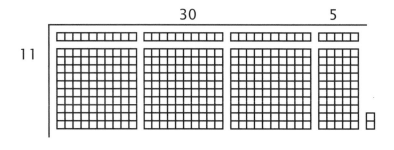

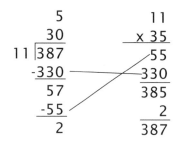

$$
\begin{array}{r}
5 \\
30 \\
11\overline{)387} \\
-330 \\
\hline
57 \\
-55 \\
\hline
2
\end{array}
\qquad
\begin{array}{r}
11 \\
\times 35 \\
\hline
55 \\
330 \\
\hline
385 \\
2 \\
\hline
387
\end{array}
$$

Example 4

Solve 303 ÷13.

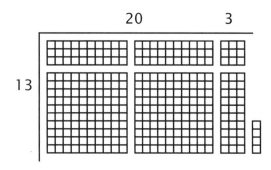

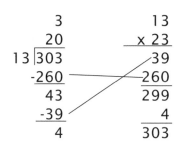

$$
\begin{array}{r}
3 \\
20 \\
13\overline{)303} \\
-260 \\
\hline
43 \\
-39 \\
\hline
4
\end{array}
\qquad
\begin{array}{r}
13 \\
\times 23 \\
\hline
39 \\
260 \\
\hline
299 \\
4 \\
\hline
303
\end{array}
$$

Hint: Start with three hundreds and three units. Break up one of the hundreds into 10 tens.

Now we'll try some without the blocks and work them through carefully. Begin by estimating the answer.

Example 5

```
    12      61  11      (70)
  × 60   12⎯743  ⎯⎯   ⎯⎯⎯⎯   What is the largest tens multiple
   720      -720  12  (10)⎤(700)  times 12 that goes into 743?
              23               The answer is 60.
    12        -12
  ×  1         11     Multiply: 12 x 60 = 720.
    12                Place 720 under the 743 and subtract. 743
                      minus 720 leaves 23.

                 12   Now what is the largest units multiple times
    61  11      × 61   12 that goes into 23? It is 1.
12 ⎯743  ⎯⎯     ⎯⎯   Multiply: 12 x 1 = 12.
   -720   12      12   Place 12 under the 23 and subtract.
     23         720   The remainder is 11.
    -12         732   We can also write the remainder as 11/12.
     11       r. 11
               743
```

When you were finding the first factor, you may have run into some difficulty. Here are some tips.

1. Use estimation. Cover the 3 with your mitten, so you have "What times 12 is 74?" You can even move the mitten over to see "What times 1 is 7?" (See figure 1.) The answer is 7.

Figure 1

```
                  6🖐            7🖐
12⎯74🖐   12⎯74🖐   12⎯74🖐   12⎯74🖐
```

2. Experiment. If you multiply 7 times 12, you will get 84, which is too large. So then we try 6 and find 6 x 12 is 72, and 60 x12 is 720, which is just right. It is perfectly normal to experiment until you get it right.

Figure 2

```
           70        60
       12|743    12|743
          840       720
       too much   just right
```

Example 6

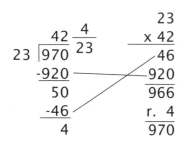

```
  23        42  4
× 40     23|970  23
 920       -920        (40)
            50     (20)|(900)
  23       -46
·× 2         4
  46
```

What is the largest tens multiple times 23 that goes into 970? It is 40.
Multiply: 23 × 40 = 920.
Place 920 under the 970 and subtract.
970 minus 920 leaves 50.

```
         42  4        23
     23|970  23     × 42
       -920           46
         50          920
        -46          966
          4        r. 4
                    970
```

Now what is the largest units multiple times 23 that goes into 50? It is 2.
Multiply: 23 × 2 = 46.
Place 46 under the 50 and subtract.
The remainder is 4.
We can also write the remainder as 4/23.

Division, Four Digit by One Digit

In this lesson, we'll learn how to divide with larger numbers. That is all that is different from what we have been doing. Take your time and study the examples, and you will be fine. When estimating, focus on the place value of the first number in the quotient. This is more important than the actual number.

Example 1

$$
\begin{array}{r}
9 \\
\times\,600 \\
\hline
5,400 \\
9 \\
\times\,70 \\
\hline
630 \\
9 \\
\times\,3 \\
\hline
27
\end{array}
\qquad
\begin{array}{r}
673\tfrac{3}{9} \\
9\,\overline{)6,060} \\
-5,400 \\
\hline
660 \\
-630 \\
\hline
30 \\
-27 \\
\hline
3
\end{array}
\qquad
\begin{array}{r}
(600) \\
(9)\overline{)(6,000)}
\end{array}
$$

What is the largest hundreds multiple times 9 that goes into 6,060? The answer is 600.
Multiply: 9 x 600 = 5,400.
Subtract from 6,060.
This leaves 660.

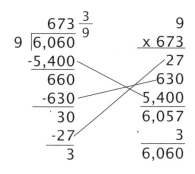

What is the largest tens multiple times 9 that goes into 660? It is 70.
Multiply: 9 x 70 = 630. Subtract from 660.
This leaves 30.

What is the largest units multiple times 9 that goes into 30? It is 3.
Multiply: 9 x 3 = 27. Subtract from 30.
This leaves 3.

Example 2

$$\begin{array}{r} 2{,}127\tfrac{1}{4} \\ 4\,\overline{)8{,}509} \\ -8{,}000 \\ \hline 509 \\ -400 \\ \hline 109 \\ -80 \\ \hline 29 \\ -28 \\ \hline 1 \end{array}$$

$$\begin{array}{r} 4 \\ \times 2{,}000 \\ \hline 8{,}000 \end{array}$$

$(2{,}000)$

$(4)\,\overline{)(9{,}000)}$

What is the largest thousands multiple times 4 that goes into 8,509? It is 2,000.
Multiply: 4 x 2,000 = 8,000.
Subtract from 8,509.
This leaves 509.

$$\begin{array}{r} 4 \\ \times 100 \\ \hline 400 \end{array}$$

What is the largest hundreds multiple times 4 that goes into 509? It is 100.
Multiply: 4 x 100 = 400.
Subtract from 509.
This leaves 109.

$$\begin{array}{r} 4 \\ \times 20 \\ \hline 80 \end{array}$$

What is the largest tens multiple times 4 that goes into 109? It is 20.
Multiply: 4 x 20 = 80.
Subtract from 109. This leaves 29.

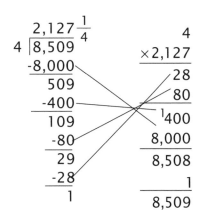

$$\begin{array}{r} 4 \\ \times 2{,}127 \\ \hline 28 \\ 80 \\ {}^{1}400 \\ 8{,}000 \\ \hline 8{,}508 \\ 1 \\ \hline 8{,}509 \end{array}$$

What is the largest units multiple times 4 that goes into 29? It is 7.
Multiply: 4 x 7 = 28.
Subtract from 29. This leaves 1.

Division, Four Digit by Two Digit

In this lesson, we'll learn how to divide larger numbers with double-digit factors. This is just one more step in getting the student to be able to divide almost any number. It is essential that students possess the ability to work through complex problems, but there does come a point when we are grateful for calculators.

Example 1

$$
\begin{array}{r}
395\frac{11}{12} \\
12\overline{)4,751} \\
-3,600 \\
\hline
1,151 \\
-1,080 \\
\hline
71 \\
-60 \\
\hline
11
\end{array}
$$

$$
\begin{array}{r}
12 \\
\times\,300 \\
\hline
3,600
\end{array}
$$

$$
\begin{array}{r}
(500) \\
(10)\overline{)(5,000)}
\end{array}
$$

What is the largest hundreds multiple times 12 that goes into 4,751? It is 300. Multiply: 12 x 300 = 3,600.

Subtract from 4,751. This leaves 1,151.

$$
\begin{array}{r}
12 \\
\times\quad 90 \\
\hline
1 \\
980 \\
\hline
1,080
\end{array}
$$

What is the largest tens multiple times 12 that goes into 1,151? It is 90. Multiply: 12 x 90 = 1,080. Subtract from 1,151. This leaves 71.

$$
\begin{array}{r}
12 \\
\times\quad 5 \\
\hline
1 \\
50 \\
\hline
60
\end{array}
$$

What is the largest units multiple times 12 that goes into 71? It is 5.

Multiply: 12 x 5 = 60. Subtract from 71. This leaves a remainder of 11 or 11/12.

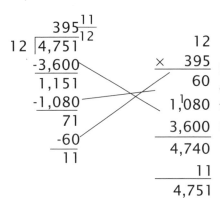

$$\begin{array}{r} 395\tfrac{11}{12} \\ 12\,\overline{)4{,}751} \\ -3{,}600 \\ \hline 1{,}151 \\ -1{,}080 \\ \hline 71 \\ -60 \\ \hline 11 \end{array}$$

$$\begin{array}{r} 12 \\ \times\ 395 \\ \hline 60 \\ 1{,}080 \\ 3{,}600 \\ \hline 4{,}740 \\ 11 \\ \hline 4{,}751 \end{array}$$

Because we had already worked these out on our worksheet to the left of the problem, I took some shortcuts in multiplying.

Example 2

$$\begin{array}{r} 38 \\ \times\ 200 \\ \hline 8{,}000 \end{array}$$

$$\begin{array}{r} 238\tfrac{19}{38} \\ 38\,\overline{)9{,}063} \\ -7{,}600 \\ \hline 1{,}463 \\ -1{,}140 \\ \hline 323 \\ -304 \\ \hline 19 \end{array}$$

$$\begin{array}{r} (200) \\ (400)\,\overline{)(9{,}000)} \end{array}$$

A good place to begin finding the first factor is with the results of estimating. So we begin by trying 200.

$$\begin{array}{r} 38 \\ \times\ 30 \\ \hline 1{,}140 \end{array}$$

What is the largest hundreds multiple times 38 that goes into 9,063? It is 200. Multiply: 38 x 200 = 7,600. Subtract from 9,063. This leaves 1,463.

$$\begin{array}{r} 38 \\ \times\ 8 \\ \hline 304 \end{array}$$

What is the largest tens multiple times 38 that goes into 1,463? It is 30. Multiply: 38 x 30 = 1,140. Subtract from 1,463. This leaves 323.

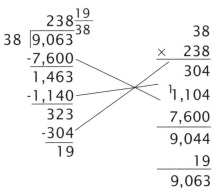

$$\begin{array}{r} 238\tfrac{19}{38} \\ 38\,\overline{)9{,}063} \\ -7{,}600 \\ \hline 1{,}463 \\ -1{,}140 \\ \hline 323 \\ -304 \\ \hline 19 \end{array}$$

$$\begin{array}{r} 38 \\ \times\ 238 \\ \hline 304 \\ 1{,}104 \\ 7{,}600 \\ \hline 9{,}044 \\ 19 \\ \hline 9{,}063 \end{array}$$

What is the largest units multiple times 38 that goes into 323? It is 8. Multiply: 38 x 8 = 304. Subtract from 323. This leaves a remainder of 19 or 19/38.

Mental Math

Here are some more mental math problems for you to read aloud to your student. You may shorten these if he or she is not ready for ones of this length.

1. Twenty-five divided by five plus three times seven equals ? (56)

2. Sixteen minus nine times two plus one equals ? (15)

3. Two plus three times four divided by two equals ? (10)

4. Five times six divided by ten times six ? (18)

5. Twenty-five minus one divided by eight plus four equals ? (7)

6. Seventy-two divided by nine minus one times six equals ? (42)

7. Forty-four plus one divided by five divided by three equals ? (3)

8. Six minus five plus eleven divided by six equals ? (2)

9. Two times seven plus two divided by two equals ? (8)

10. Seven times seven minus one divided by eight times five equals ? (30)

DIVISION, FOUR DIGIT BY TWO DIGIT - LESSON 24   **89**

Division, Multiple Digit by Multiple Digit

By now someone has probably talked about how to write fewer numbers in working out the answer. So in examples 1 and 2, I show the shortened version below the primary version. Study this and use it if you understand that they are still the same numbers, with the same values, just abbreviated. Notice that after you subtract, you bring down only one additional number (or one place value) before dividing again.

Example 1

```
        53
    ×  500                    538 5/53
        1       53 )30,109
    25,500           -26,500
    26,500             3,609
                      -3,180
        53               429
    ×   60               424
        1                  5
     3,080
     3,180

        53
    ×    8
         2
       404
       424
```

$$(50) \overline{)(30,000)}^{(600)}$$

What is the largest thousands multiple times 53 that goes into 30,109? Is it 1,000? No, because 53 × 1,000 = 53,000, which is larger than 30,109.

What is the largest hundreds multiple times 53 that goes into 30,109? Is it 600, which is our estimate? No, because 600 times 53 is 31,800, which is still too large. Try 500. Multiply 53 × 500 = 26,500 and subtract from 30,109. This leaves 3,609.

What is the largest tens multiple times 53 that goes into 3,609? It is 60. Multiply 53 × 60 = 3,280 and subtract from 3,609, leaving 429.

What is the largest units multiple times 53 that goes into 429? It is 8.
Multiply 53 × 8 = 424 and subtract from 429. This leaves a remainder of 5 or 5/53.

$538 \frac{5}{53}$

53 | 30,109
-26 5↓
3 60
-3 18↓
429
-424
5

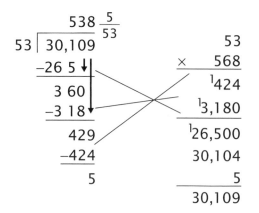

53
× 568
¹424
¹3,180
¹26,500
30,104
5
30,109

Because we had already worked these out on our worksheet to the left of the problem, I took some shortcuts in multiplying.

Example 2

136
× 4,000
12
424,000
544,000

$4,257 \frac{10}{136}$

136 | 578,962
-544,000
34,962
-27,200
7,762
-6,800
962
-952
10

(6,000)
(100) | (300,000)

What is the largest thou sands multiple times 136 that goes into 578,962? Is it 6,000? No, too big. 5,000? No, still too large. Multiply 136 x 4,000 = 544,000 and subtract.

What is the largest hundreds multiple times 136 that goes into 34,962? It is 200. Multiply 136 x 200 = 27,200 and subtract, leaving 7,762.

136
× 200
1
26,200
27,200

What is the largest tens multiple times 136 that goes 7,762? It is 50. Multiply 136 x 50 = 6,800 and subtract, leaving 962.

136
× 50
1 3
5,500
6,800

136
× 7
24
712
952

What is the largest units multiple times 136 that goes into 962? It is 7. Multiply 136 x 7 = 952 and subtract from 962. This leaves a remainder of 10 or 10/136.

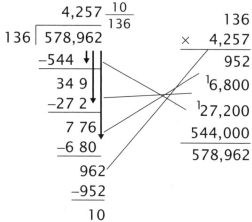

$4,257 \frac{10}{136}$

136 | 578,962
-544↓
34 9
-27 2↓
7 76
-6 80
962
-952
10

136
× 4,257
952
¹6,800
¹27,200
544,000
578,962

LESSON 26

Volume

When doing these problems, I relate our figure to a hotel. First, find the number of rooms on the ground floor, and then multiply times the number of floors. The formula is V = Bh. (Read this as *volume* is the area of the base times the height.) The capital *B* is the area of the base or the number of rooms on the first floor. The height is the number of floors in the hotel. As in all the formulas where *h* represents height, such as for a rectangle, the height is perpendicular to the base, or in this case, the *B*.

In the first few problems, when we introduce volume, the cubes are clearly shown, so have the students build the problem with a stack of blocks. After this is done, write in the dimensions of the figure. In subsequent problems, the dimensions will be given, as in example 3, but the blocks, or cubic units, will not be shown. Of course, the student may draw them in if that would be helpful. Remember, the units of measure in a volume problem are cubic!

Example 1
Find the volume.

V = Bh
V = (4 x 3) x 2
V = 24 cubic units

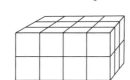

Example 2

Find the volume.

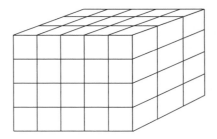

$$V = Bh$$
$$V = (5 \times 3) \times 4$$
$$V = 60 \text{ cubic units}$$

Example 3

Find the volume.

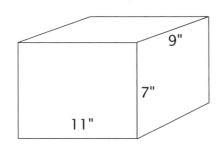

9"

7"

11"

$$V = Bh$$
$$V = (11" \times 9") \times 7"$$
$$V = 693 \text{ cubic inches (cu in)}$$

Volume and Weight of Water - In the student text, there are problems that ask the amount of water in a cubic foot. There are 7.48 gallons of water in a cubic foot, and the weight of a gallon is 8.345 pounds. So we will round the gallons to seven gallons per cubic foot and the weight to eight pounds per gallon.

Fraction of a Number

There are three steps that will help you to understand fractions. Work on these in this lesson, and we will use them as a basis for much of the study of fractions. The first step is often left out of most teaching about fractions. It is very important because it is what you begin with. A *fraction* is a "fraction of" something. This lesson magnifies this fact. I often ask students, "What is larger, one-half or one-fourth?" They normally reply "One-half," but then I say, "One-half of the room we are in or one-fourth of the state?" The correct response is "One-half of what?" and "One-fourth of what?"

The second step is determining the *denominator*. This is the number of equal parts into which you divide the starting number. There is a natural language connection, since both "divide" and "denominator" begin with a *d*. The symbolism is also an indicator of this step, since the line separating the numerator and denominator means "divided by." So you divide step one by the bottom number of the fraction —the denominator—into that many equal parts.

In step three, you count how many of the equal parts. I call this the "numBerator" at first to make the connection with counting. But after a while, we take out the *B* and have *numerator*. To summarize, here are the three points.

1. The number you start with; in example 1, this is six.
2. The denominator, which indicates the **value** (as in place value) and tells how many equal parts we divide the starting number into.
3. The "numBerator" or numerator, which is **how many** of these equal parts we count.

Example 1

Find $\frac{2}{3}$ of 6.

$\frac{2}{3}$ of 6 = 4

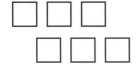

Step 1	**Step 2**	**Step 3**
Select six green unit blocks.	Divide six into three equal parts.	Count two of those parts.

A fraction is a combination of a division problem and a multiplication problem. First, you are dividing (in example 1, by three for the equal parts) to get the denominator, and then you are multiplying (by two) to get the numerator. $6 \div 3 \times 2 = 4$.

Example 2

Find $\frac{3}{4}$ of 12.

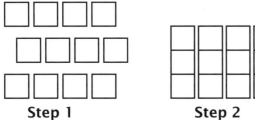

$\frac{3}{4}$ of 12 = 9

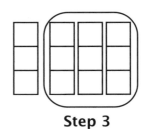

Step 1	**Step 2**	**Step 3**
Select 12 green unit blocks.	Divide 12 into four equal parts.	Count three of those parts.

Multi-Step Word Problems

Here are a few more multi-step word problems to try. You may want to read and discuss these with your student as you work out the solutions together. Remember that the purpose is to stretch, not to frustrate.

The answers are at the end of the solutions at the back of this book.

1. Sixty-five bags of nuts are to be divided among 13 students. Each bag contains 15 nuts. How many nuts will each student receive?

2. Shane is playing a board game. For his first turn, he moved ahead three spaces; for the second, five spaces; and for the third, one space. For his next turn, he had to go back six spaces. After that he got a card that said he could move two times the biggest forward move he had done so far. Now how many spaces from the beginning is Shane's game piece?

3. The volume of a rectangular box is 330 cubic inches. The length on one side of the top is 11 inches, and the height of the box is 3 inches. What is the area of the top of the box?
(A drawing may help you with this one.)

4. Tom divided $360 among his six children for them to use for Christmas gifts. His daughter Kate added $20 to her portion, and then used the money to buy 16 gifts that each cost the same amount. What was the cost of each of Kate's gifts?

Roman Numerals: I, V, X, L, and C

In this lesson and in lesson 30, we will explore the four rules and seven symbols that comprise Roman numerals. You often see these numbers representing the year that a building was constructed or that a movie was made. They also pop up occasionally in unexpected places and events, so we will learn them here. In this lesson we will introduce three of the rules and five of the symbols. A capital "I" represents 1. Capital "V" represents 5, and capital "X" represents 10. To show the number 3, you would write III. To show the number 30, you would write XXX.

Rule 1   You can't use more than three of the same letters
in a row when using I and X.

Rule 2   You can use V only once. There will never be a VVV.

Rule 3   If I and X are to the left of a larger symbol,
they are subtracted from that symbol.

Here are the numbers 1 through 39 represented with these three symbols and the three rules.

1	I	11	XI	21	XXI	31	XXXI
2	II	12	XII	22	XXII	32	XXXII
3	III	13	XIII	23	XXIII	33	XXXIII
4	IV	14	XIV	24	XXIV	34	XXXIV
5	V	15	XV	25	XXV	35	XXXV
6	VI	16	XVI	26	XXVI	36	XXXVI
7	VII	17	XVII	27	XXVII	37	XXXVII
8	VIII	18	XVIII	28	XXVIII	38	XXXVIII
9	IX	19	XIX	29	XXIX	39	XXXIX
10	X	20	XX	30	XXX		

Notice that we never use I or X more than three times in a row. Notice as well that V cannot be repeated, and that when I is to the left of a V or an X, it is subtracted. See how you form 3 using the symbol I three times. When you make the number 4, you have to use 5 minus 1, which is what the IV means, because I is to the left of V.

We had to stop at 39 because we need a larger symbol to proceed. We can only use three tens, or XXX, so we need a symbol for 50, and then we can put X to the left of it to get 40.

Roman Numerals - L and C

Now that you are familiar with I, V, and X, there are two more symbols to introduce to you. "L" represents 50 and "C" is the symbol for 100. With these new symbols, and some additions to the same rules, we can write the Roman numerals for all of the numbers from 1 to 399. Notice the modifications to rules 1 and 2.

Rule 1   You can't use more than three of the same letter in a row when using I, X, or C.

Rule 2   You can use V and L only once. There will never be a VVV or an LL.

Rule 3   If I and X are to the left of a larger symbol, they are subtracted from that symbol.

Here are the larger numbers from 40 through 350, represented with these five symbols and the three rules.

40	XL	100	C
50	L	150	CL
60	LX	200	CC
70	LXX	250	CCL
80	LXXX	300	CCC
90	XC	350	CCCL

Example 1
Show 168 with Roman numerals.

168 is 100 + 50 + 10 + 5 + 3, which is C + L + X + V + III or CLXVIII.

Example 2
Show 249 with Roman numerals.

249 is 200 + 40 + 5 + 4, which is CC + XL+ IX or CCXLIX.

Example 3
What number is represented by CXCIII?

C is 100, XC is 90, and III is 3, so the number is 193.

Example 4
What number is represented by CCCLXXIX?

CCC is 300, L is 50, XX is 20, and IX is 9,
so the number is 379.

Fraction of One

Look at the square in example 1. This represents one or one whole (the first step). The second step is to count how many equal pieces the square is divided into. This is consistent with the written problem where the number below the line is 5, and the line means "divided by." The square is divided into five equal parts. Now count the number of shaded parts. The number on top of the line tells us how many of the five equal parts are shaded. The bottom number is referred to as the denominator, and the top number is the "numberator" or numerator. There are three steps: what we start with, how many parts we divide it into, and how many of the divided parts we count or number.

Example 1

Find $\frac{2}{5}$ of 1.

 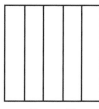 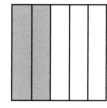

Step 1	**Step 2**	**Step 3**
Begin with one.	Divide one into five equal parts.	Count two of those parts.

Example 2

Find $\frac{3}{4}$ of 1.

 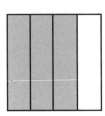 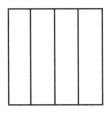

Step 1	**Step 2**	**Step 3**
Begin with one.	Divide one into four equal parts.	Count three of those parts.

Example 3

Find $\frac{2}{3}$ of 1.

 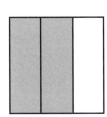 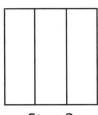

Step 1	**Step 2**	**Step 3**
Begin with one.	Divide one into three equal parts.	Count two of those parts.

Roman Numerals: D, M, and Multiples of 1,000

The only symbols left are "D" for 500 and "M" for 1,000. We also need to modify rules 1 and 2, and add rule 4. With these new symbols and rules, we can write the Roman numerals for all of the numbers from 1 to 3,999,999.

Rule 1   You can't use more than three letters in a row when using I, X, C, and M.

Rule 2   You can use V, L, and D only once. There will never be a VVV, an LL, or a DDD.

Rule 3   If I, X, or C are to the left of a larger symbol, they are subtracted from that symbol.

Rule 4   To show 1,000 times a number, put a line over the symbol.

$\overline{V}$ = 5,000, $\overline{M}$ = 1,000,000, and $\overline{D}$ = 500,000.

Study the chart of some significant numbers below, as well as the examples. Make up some yourself.

400	CD	1,000	M
500	D	4,000	$M\overline{V}$
600	DC	10,000	$\overline{X}$
700	DCC	100,000	$\overline{C}$
800	DCCC	700,000	$\overline{DCC}$
900	CM	900,000	$\overline{CM}$

Example 1
Show 243,792 with Roman numerals.

243,792 is 200,000 + 40,000 + 3,000 + 700 + 90 + 2, which is $\overline{CCXL}$MMMDCCXCII.

Example 2
Show 69,508 with Roman numerals.

69,508 is 60,000 + 9,000 + 500 + 8, which is $\overline{LXIX}$DVIII.

Example 3
What number is represented by $\overline{MMM}\,\overline{LXX}$CIV?

$\overline{MMM}$ is 3,000,000, $\overline{LXX}$ is 70,000, C is 100, and IV is 4, so the number is 3,070,104.

Example 4
What number is represented by $\overline{V}$MMMDCCXCIII?

$\overline{V}$ is 5,000, MMM is 3,000, DCC is 700, XC is 90, and III is 3, so the number is 8,793.

Student Solutions

Lesson Practice 1A

1. Done
2. $3 \times 3 = 9$
3. $6 \times 2 = 12$
   $2 \times 6 = 12$
4. $4 \times 3 = 12$
   $3 \times 4 = 12$
5. $5 \times 5 = 25$
6. $4 \times 2 = 8$
   $2 \times 4 = 8$
7. $6 \times \underline{6} = 36$
8. $2 \times \underline{10} = 20$
9. $4 \times \underline{7} = 28$
10. $5 \times \underline{4} = 20$
11. $7 \times \underline{3} = 21$
12. $8 \times \underline{3} = 24$
13. $7 \times \underline{7} = 49$
14. $5 \times \underline{6} = 30$

Lesson Practice 1B

1. $2 \times 2 = 4$
2. $6 \times 4 = 24$
   $4 \times 6 = 24$
3. $5 \times 2 = 10$
   $2 \times 5 = 10$
4. $5 \times 4 = 20$
   $4 \times 5 = 20$
5. $7 \times 4 = 28$
   $4 \times 7 = 28$
6. $3 \times 2 = 6$
   $2 \times 3 = 6$
7. $3 \times \underline{1} = 3$
8. $2 \times \underline{2} = 4$
9. $1 \times \underline{10} = 10$
10. $8 \times \underline{2} = 16$
11. $7 \times \underline{2} = 14$
12. $5 \times \underline{6} = 30$

13. $10 \times \underline{10} = 100$
14. $5 \times \underline{9} = 45$

Lesson Practice 1C

1. $7 \times 2 = 14$
   $2 \times 7 = 14$
2. $5 \times 3 = 15$
   $3 \times 5 = 15$
3. $7 \times 3 = 21$
   $3 \times 7 = 21$
4. $4 \times 4 = 16$
5. $8 \times 1 = 8$
   $1 \times 8 = 8$
6. $3 \times 4 = 12$
   $4 \times 3 = 12$
7. $8 \times \underline{7} = 56$
8. $9 \times \underline{4} = 36$
9. $6 \times \underline{4} = 24$
10. $3 \times \underline{9} = 27$
11. $5 \times \underline{9} = 45$
12. $6 \times \underline{3} = 18$
13. $10 \times \underline{9} = 90$
14. $2 \times \underline{8} = 16$

Systematic Review 1D

1. $4 \times 2 = 8$
   $2 \times 4 = 8$
2. $5 \times 6 = 30$
   $6 \times 5 = 30$
3. $7 \times \underline{7} = 49$
4. $10 \times \underline{8} = 80$
5. $3 \times \underline{8} = 24$
6. $4 \times \underline{9} = 36$

7. $2 \times 4 = 8$ sq ft

8. $2 \times 2 = 4$ sq ft

Systematic Review 1E

1. $6 \times 6 = 36$
2. $4 \times 7 = 28$
   $7 \times 4 = 28$
3. $9 \times \underline{9} = 81$
4. $3 \times \underline{5} = 15$
5. $7 \times \underline{6} = 42$
6. $2 \times \underline{5} = 10$
7. $10 \times 10 = 100$ sq mi
8. $8 \times 4 = 32$ sq in
9. $4 \times 5 = 20$ sq ft
10. $4F = 20;\ F = 5$
11. $3H = 30;\ H = 10$
12. $3 \times 2 = 6$ sq ft

Systematic Review 1F

1. $3 \times 5 = 15$
   $5 \times 3 = 15$
2. $2 \times 6 = 12$
   $6 \times 2 = 12$
3. $5 \times \underline{5} = 25$
4. $7 \times \underline{8} = 56$
5. $9 \times \underline{6} = 54$
6. $3 \times \underline{6} = 18$
7. $6 \times 6 = 36$ sq ft
8. $8 \times 5 = 40$ sq in
9. $7 \times 6 = 42$ sq mi
10. $4D = 24;\ D = 6$
11. $9H = 36;\ H = 4$
12. $6 \times 5 = 30$ sq mi

Lesson Practice 2A

1. $2,4,6,8;\underline{4}$
2. $1,2,3,4,5,6,7,8,9,10;\underline{10}$
3. $2,4,6,8,10,12,14;\underline{7}$
4. $2,4,6;\underline{3}$
5. $1,2,3,4,5,6,7;\underline{7}$
6. $16 \div 2 = \underline{8}$
7. $9 \div 1 = \underline{9}$
8. $4 \div 2 = \underline{2}$
9. $5 \div 1 = \underline{5}$
10. $18 \div 2 = \underline{9}$
11. $8 \div 1 = \underline{8}$
12. $12 \div 2 = \underline{6}$
13. $2 \div 1 = \underline{2}$
14. $4 \div 1 = \underline{4}$
15. $\dfrac{10}{2} = \underline{5}$
16. $\dfrac{2}{2} = \underline{1}$
17. $\dfrac{6}{1} = \underline{6}$
18. $\dfrac{1}{1} = \underline{1}$
19. $\dfrac{14}{2} = \underline{7}$
20. $\dfrac{8}{2} = \underline{4}$
21. $16 \div 2 = 8$ people
22. $10 \div 2 = 5$ cookies

Lesson Practice 2B

1. $2,4,6,8,10;\underline{5}$
2. $2,4,6,8,10,12,14,16;\underline{8}$
3. $1;\underline{1}$
4. $1,2,3,4,5;\underline{5}$
5. $2,4,6,8,10,12;\underline{6}$
6. $10 \div 2 = \underline{5}$
7. $16 \div 2 = \underline{8}$
8. $14 \div 2 = \underline{7}$
9. $2 \div 2 = 1$
10. $2 \div 1 = \underline{2}$

11. $18 \div 2 = \underline{9}$

12. $9 \div 1 = \underline{9}$

13. $8 \div 2 = \underline{4}$

14. $6 \div 1 = \underline{6}$

15. $\dfrac{4}{1} = \underline{4}$

16. $\dfrac{8}{1} = \underline{8}$

17. $\dfrac{4}{2} = \underline{2}$

18. $\dfrac{7}{1} = \underline{7}$

19. $\dfrac{6}{2} = \underline{3}$

20. $\dfrac{14}{2} = \underline{7}$

21. $8 \div 1 = 8$ people

22. $8 \div 2 = 4$ people

Lesson Practice 2C

1. $1,2,3,4,5,6,7,8,9;\underline{9}$

2. $2,4,6,8,10,12,14,16,18;\underline{9}$

3. $1,2,3,4;\underline{4}$

4. $2,4,6,8,10,12,14,16,18,20;\underline{10}$

5. $2,4;\underline{2}$

6. $2 \div 2 = \underline{1}$

7. $8 \div 2 = \underline{4}$

8. $6 \div 2 = \underline{3}$

9. $4 \div 2 = \underline{2}$

10. $6 \div 1 = \underline{6}$

11. $12 \div 2 = \underline{6}$

12. $10 \div 1 = \underline{10}$

13. $16 \div 2 = \underline{8}$

14. $7 \div 1 = \underline{7}$

15. $\dfrac{14}{2} = \underline{7}$

16. $\dfrac{1}{1} = \underline{1}$

17. $\dfrac{18}{2} = \underline{9}$

18. $\dfrac{9}{1} = \underline{9}$

19. $\dfrac{20}{2} = \underline{10}$

20. $\dfrac{10}{2} = \underline{5}$

21. $12 \div 2 = 6$ people

22. $6 \div 1 = 6$ people

Systematic Review 2D

1. $4 \div 2 = \underline{2}$

2. $8 \div 1 = \underline{8}$

3. $4 \div 1 = \underline{4}$

4. $18 \div 2 = \underline{9}$

5. $2 \div 2 = \underline{1}$

6. $2 \div 1 = \underline{2}$

7. $\dfrac{14}{2} = \underline{7}$

8. $\dfrac{16}{2} = \underline{8}$

9. $\dfrac{5}{1} = \underline{5}$

10. $\begin{array}{r} 5 \\ \times\ 6 \\ \hline 30 \end{array}$

11. $\begin{array}{r} 10 \\ \times\ 9 \\ \hline 90 \end{array}$

12. $10 \times 6 = 60$

13. $(5)(7) = 35$

14. $5 \times \underline{7} = 35$

15. $5 \times \underline{9} = 45$

16. $10 \times \underline{1} = 10$

17. $10 \times \underline{4} = 40$

18. $5 \times 5 = \underline{25}$ sq ft

19. $5 \times 3 = \underline{15}$ sq ft

20. $3 \times 10 = \underline{30}$ sq ft

21. $18 \div 2 = 9$

22. $18 \div 2 = 9$ piles

Systematic Review 2E

1. $6 \div 1 = \underline{6}$
2. $8 \div 2 = \underline{4}$
3. $9 \div 1 = \underline{9}$
4. $18 \div 2 = \underline{9}$
5. $2 \div 2 = \underline{1}$
6. $2 \div 1 = \underline{2}$
7. $\dfrac{12}{1} = \underline{12}$
8. $\dfrac{10}{2} = \underline{5}$
9. $\dfrac{1}{1} = \underline{1}$
10. $5 \times 1 = 5$
11. $8 \times \underline{10} = 80$
12. $7 \times \underline{5} = 35$
13. $3 \times \underline{10} = 30$
14. $\begin{array}{r} 8 \\ \times\ 5 \\ \hline 40 \end{array}$
15. $\begin{array}{r} 10 \\ \times\ 7 \\ \hline 70 \end{array}$
16. $2 \times 10 = \underline{20}$
17. $10 \cdot 10 = \underline{100}$
18. $2 \times 2 = \underline{4}$ sq ft
19. $5 \times 10 = \underline{50}$ sq mi
20. $3 \times 5 = \underline{15}$ sq in
21. $8 \div 1 = 8$
22. $20 \div 2 = 10$ cages

Systematic Review 2F

1. $5 \div 1 = \underline{5}$
2. $16 \div 2 = \underline{8}$
3. $4 \div 2 = \underline{2}$
4. $8 \div 1 = \underline{8}$
5. $14 \div 2 = \underline{7}$
6. $12 \div 2 = \underline{6}$

7. $\dfrac{18}{2} = \underline{9}$
8. $\dfrac{10}{2} = \underline{5}$
9. $\dfrac{20}{2} = \underline{10}$
10. $5 \times \underline{4} = 20$
11. $5 \times \underline{10} = 50$
12. $9 \times \underline{10} = 90$
13. $5 \times \underline{8} = 40$
14. $\begin{array}{r} 9 \\ \times\ 5 \\ \hline 45 \end{array}$
15. $\begin{array}{r} 10 \\ \times\ 4 \\ \hline 40 \end{array}$
16. $2 \cdot 5 = \underline{10}$
17. $(10)(8) = \underline{80}$
18. $10 \times 10 = 100$ sq in
19. $1 \times 1 = 1$ sq mi
20. $6 \times 5 = 30$ sq ft
21. $2 \div 2 = 1$
22. $14 \div 2 = 7$ stickers

Lesson Practice 3A

1. $10,20,30,40,50,60,70,80; \underline{8}$
2. $10,20,30,40,50,60,70,80,90,100; \underline{10}$
3. $10,20,30,40,50,60; \underline{6}$
4. $10; \underline{1}$
5. $50 \div 10 = \underline{5}$
6. $100 \div 10 = \underline{10}$
7. $30 \div 10 = \underline{3}$
8. $90 \div 10 = \underline{9}$
9. $20 \div 10 = \underline{2}$
10. $40 \div 10 = \underline{4}$
11. $10 \div 10 = \underline{1}$
12. $70 \div 10 = \underline{7}$
13. $50 \div 10 = \underline{5}$
14. $\dfrac{30}{10} = \underline{3}$

15. $\dfrac{90}{10} = \underline{9}$

16. $\dfrac{60}{10} = \underline{6}$

17. $80 \div 10 = 8$ teams

18. $\$40 \div \$10 = 4$ books

Lesson Practice 3B

1. 10,20;$\underline{2}$
2. 10,20,30,40;$\underline{4}$
3. 10,20,30,40,50,60,70,80,90;$\underline{9}$
4. 10,20,30;$\underline{3}$
5. $60 \div 10 = \underline{6}$
6. $90 \div 10 = \underline{9}$
7. $50 \div 10 = \underline{5}$
8. $30 \div 10 = \underline{3}$
9. $70 \div 10 = \underline{7}$
10. $10 \div 10 = \underline{1}$
11. $40 \div 10 = \underline{4}$
12. $20 \div 10 = \underline{2}$
13. $90 \div 10 = \underline{9}$
14. $\dfrac{100}{10} = \underline{10}$
15. $\dfrac{50}{10} = \underline{5}$
16. $\dfrac{10}{10} = \underline{1}$
17. $70 \div 7 = 7$ jelly beans
18. $30 \div 10 = 3$ sheets

Lesson Practice 3C

1. 10,20,30,40,50;$\underline{5}$
2. 10,20,30,40,50,60,70;$\underline{7}$
3. 10,20,30,40,50,60,70,80,90,100;$\underline{10}$
4. 10,20,30,40,50,60;$\underline{6}$
5. $90 \div 10 = \underline{9}$
6. $40 \div 10 = \underline{4}$
7. $100 \div 10 = \underline{10}$
8. $50 \div 10 = \underline{5}$
9. $80 \div 10 = \underline{8}$

10. $60 \div 10 = \underline{6}$
11. $20 \div 10 = \underline{2}$
12. $30 \div 10 = \underline{3}$
13. $10 \div 10 = \underline{1}$
14. $\dfrac{70}{10} = \underline{7}$
15. $\dfrac{40}{10} = \underline{4}$
16. $\dfrac{30}{10} = \underline{3}$
17. $50 \div 10 = 5$ hours
18. $100 \div 10 = 10$ pennies

Systematic Review 3D

1. $10 \div 10 = \underline{1}$
2. $30 \div 10 = \underline{3}$
3. $60 \div 10 = \underline{6}$
4. $70 \div 10 = \underline{7}$
5. $5 \div 1 = \underline{5}$
6. $14 \div 2 = \underline{7}$
7. $100 \div 10 = \underline{10}$
8. $10 \div 1 = \underline{10}$
9. $18 \div 2 = \underline{9}$
10. $8 \div 1 = \underline{8}$

11. $\dfrac{50}{10} = \underline{5}$
12. $\dfrac{16}{2} = \underline{8}$
13. $\begin{array}{r} 9 \\ \times\ 3 \\ \hline 27 \end{array}$
14. $\begin{array}{r} 3 \\ \times\ 4 \\ \hline 12 \end{array}$
15. $3 \times 7 = \underline{21}$
16. $10 \cdot 3 = \underline{30}$
17. $12 \div 2 = 6$ qt
18. $8 \times 3 = 24$ ft
19. $5 \times 8 = 40$ sq in
20. $12 \div 2 = 6$ cans

Systematic Review 3E

1. $20 \div 10 = \underline{2}$
2. $40 \div 10 = \underline{4}$
3. $90 \div 10 = \underline{9}$
4. $80 \div 10 = \underline{8}$
5. $7 \div 1 = \underline{7}$
6. $20 \div 2 = \underline{10}$
7. $10 \div 10 = \underline{1}$
8. $6 \div 1 = \underline{6}$
9. $8 \div 2 = \underline{4}$
10. $4 \div 1 = \underline{4}$
11. $\dfrac{6}{2} = \underline{3}$
12. $\dfrac{14}{2} = \underline{7}$
13. $\begin{array}{r} 3 \\ \times 3 \\ \hline 9 \end{array}$
14. $\begin{array}{r} 5 \\ \times\ 3 \\ \hline 15 \end{array}$
15. $(2)(3) = \underline{6}$
16. $6 \times 3 = \underline{18}$
17. $14 \div 2 = 7$ qt
18. $5 \times 3 = 15$ ft
19. $1 \times 2 = 2$ sq mi
20. $18 \div 2 = 9$ people

Systematic Review 3F

1. $70 \div 10 = \underline{7}$
2. $60 \div 10 = \underline{6}$
3. $50 \div 10 = \underline{5}$
4. $30 \div 10 = \underline{3}$
5. $9 \div 1 = \underline{9}$
6. $16 \div 2 = \underline{8}$
7. $4 \div 2 = \underline{2}$
8. $3 \div 1 = \underline{3}$
9. $10 \div 2 = \underline{5}$
10. $12 \div 2 = \underline{6}$

11. $\dfrac{2}{2} = \underline{1}$
12. $\dfrac{18}{2} = \underline{9}$
13. $\begin{array}{r} 4 \\ \times\ 3 \\ \hline 12 \end{array}$
14. $\begin{array}{r} 3 \\ \times\ 8 \\ \hline 24 \end{array}$
15. $7 \cdot 3 = \underline{21}$
16. $3 \times 9 = \underline{27}$
17. $16 \div 2 = 8$ qt
18. $6 \times 3 = 18$ sq yd
19. $6 \times 3 = 18$ ft long
    $3 \times 3 = 9$ ft wide
20. $3 \times 10 = 30$ cookies
    $30 \div 10 = 3$ cookies

Lesson Practice 4A

1. $5,10,15,20,25,30,35;\underline{7}$
2. $5,10,15,20,25,30,35,40,45,50;\underline{10}$
3. $5;\underline{1}$
4. $3,6,9,12,15,18;\underline{6}$
5. $45 \div 5 = \underline{9}$
6. $20 \div 5 = \underline{4}$
7. $15 \div 5 = \underline{3}$
8. $30 \div 3 = \underline{10}$
9. $12 \div 3 = \underline{4}$
10. $27 \div 3 = \underline{9}$
11. $30 \div 5 = \underline{6}$
12. $10 \div 5 = \underline{2}$
13. $40 \div 5 = \underline{8}$
14. $\dfrac{25}{5} = \underline{5}$
15. $\dfrac{24}{3} = \underline{8}$
16. $\dfrac{21}{3} = \underline{7}$
17. $9 \div 3 = 3$ pieces
18. $15 \div 3 = 5$ cages

Lesson Practice 4B

1. 5,10,15,20,25,30,35,40; $\underline{8}$
2. 3; $\underline{1}$
3. 5,10,15,20,25; $\underline{5}$
4. 3,6,9,12,15,18,21,24,27; $\underline{9}$
5. $5 \div 5 = \underline{1}$
6. $35 \div 5 = \underline{7}$
7. $50 \div 5 = \underline{10}$
8. $6 \div 3 = \underline{2}$
9. $24 \div 3 = \underline{8}$
10. $18 \div 3 = \underline{6}$
11. $20 \div 5 = \underline{4}$
12. $45 \div 5 = \underline{9}$
13. $21 \div 3 = \underline{7}$
14. $\frac{15}{3} = \underline{5}$
15. $\frac{30}{3} = \underline{10}$
16. $\frac{30}{5} = \underline{6}$
17. $15 \div 5 = 3$ five dollar bills
18. $21 \div 3 = 7$ apples

Lesson Practice 4C

1. 3,6,9; $\underline{3}$
2. 3,6,9,12,15,18,21,24; $\underline{8}$
3. 5,10; $\underline{2}$
4. 5,10,15,20; $\underline{4}$
5. $12 \div 3 = \underline{4}$
6. $27 \div 3 = \underline{9}$
7. $40 \div 5 = \underline{8}$
8. $15 \div 3 = \underline{5}$
9. $15 \div 5 = \underline{3}$
10. $30 \div 3 = \underline{10}$
11. $25 \div 5 = \underline{5}$
12. $50 \div 5 = \underline{10}$
13. $18 \div 3 = \underline{6}$
14. $\frac{30}{5} = \underline{6}$
15. $\frac{21}{3} = \underline{7}$

16. $\frac{5}{5} = \underline{1}$
17. $24 \div 3 = 8$ pieces
18. $35 \div 5 = 7$ hands

Systematic Review 4D

1. $6 \div 3 = \underline{2}$
2. $30 \div 5 = \underline{6}$
3. $18 \div 3 = \underline{6}$
4. $45 \div 5 = \underline{9}$
5. $15 \div 3 = \underline{5}$
6. $25 \div 5 = \underline{5}$
7. $40 \div 10 = \underline{4}$
8. $16 \div 2 = \underline{8}$
9. $24 \div 3 = \underline{8}$
10. $30 \div 3 = \underline{10}$
11. $\frac{14}{2} = \underline{7}$
12. $\frac{40}{5} = \underline{8}$
13. $4 \times \underline{3} = 12$
14. $6 \times \underline{5} = 30$
15. $5 \times \underline{3} = 15$
16. $8 \times \underline{3} = 24$
17. $\begin{array}{r} 25 \\ +34 \\ \hline 59 \end{array}$
18. $\begin{array}{r} {\scriptstyle 1} \\ 78 \\ +34 \\ \hline 112 \end{array}$
19. $\begin{array}{r} {\scriptstyle 1} \\ 49 \\ +51 \\ \hline 100 \end{array}$
20. $\begin{array}{r} {\scriptstyle 1} \\ 65 \\ +15 \\ \hline 80 \end{array}$
21. $\$39 + \$28 = \$67$
22. $\$50 \div \$5 = 10$ days

Systematic Review 4E

1. $12 \div 3 = \underline{4}$
2. $35 \div 5 = \underline{7}$
3. $15 \div 5 = \underline{3}$
4. $9 \div 3 = \underline{3}$
5. $21 \div 3 = \underline{7}$
6. $10 \div 2 = \underline{5}$
7. $27 \div 3 = \underline{9}$
8. $60 \div 10 = \underline{6}$
9. $8 \div 2 = \underline{4}$
10. $3 \div 1 = \underline{3}$
11. $\dfrac{50}{5} = \underline{10}$
12. $\dfrac{3}{3} = \underline{1}$
13. $5 \times \underline{7} = 35$
14. $3 \times \underline{7} = 21$
15. $7 \times \underline{10} = 70$
16. $6 \times \underline{3} = 18$
17. $\begin{array}{r} {\scriptstyle 1} \\ 13 \\ +19 \\ \hline 32 \end{array}$
18. $\begin{array}{r} {\scriptstyle 1} \\ 28 \\ +49 \\ \hline 77 \end{array}$
19. $\begin{array}{r} 26 \\ +72 \\ \hline 98 \end{array}$
20. $\begin{array}{r} {\scriptstyle 1} \\ 47 \\ +38 \\ \hline 85 \end{array}$
21. $20 \div 5 = 4$ stickers
22. $27 \div 3 = 9$ pages
23. $10 \times 2 = 20$ pints
    $20 + 3 = 23$ pints
24. $7 \times 2 = 14$ notes
    $14 + 11 = 25$ people thanked

Systematic Review 4F

1. $10 \div 5 = \underline{2}$
2. $24 \div 3 = \underline{8}$
3. $30 \div 3 = \underline{10}$
4. $40 \div 5 = \underline{8}$
5. $35 \div 5 = \underline{7}$
6. $15 \div 5 = \underline{3}$
7. $12 \div 3 = \underline{4}$
8. $18 \div 3 = \underline{6}$
9. $18 \div 2 = \underline{9}$
10. $70 \div 10 = \underline{7}$
11. $\dfrac{27}{3} = \underline{9}$
12. $\dfrac{5}{1} = \underline{5}$
13. $5 \times \underline{10} = 50$
14. $3 \times \underline{3} = 9$
15. $10 \times \underline{9} = 90$
16. $9 \times \underline{1} = 9$
17. $\begin{array}{r} 81 \\ +18 \\ \hline 99 \end{array}$
18. $\begin{array}{r} {\scriptstyle 1} \\ 37 \\ +37 \\ \hline 74 \end{array}$
19. $\begin{array}{r} {\scriptstyle 1} \\ 42 \\ +29 \\ \hline 71 \end{array}$
20. $\begin{array}{r} 74 \\ +25 \\ \hline 99 \end{array}$
21. $9 \times 5 = 45$ books
22. $3 \times \$5 = \$15$
    $\$15 \div 5 = \$3$ per person
23. $18 \div 3 = 6$ pieces
24. $46 + 28 = 74$ mi

Lesson Practice 5A

1. true
2. true
3. false
4. false
5. $\perp$
6. $\parallel$
7. no
8. no
9. yes
10. 2
11. perpendicular
12. parallel

Lesson Practice 5B

1. false
2. false
3. true
4. true
5. $\parallel$
6. $\perp$
7. perpendicular
8. no
9. yes
10. no
11. parallel
12. perpendicular

Lesson Practice 5C

1. true
2. true
3. false
4. false
5. $\perp$
6. $\parallel$
7. no
8. no

9. yes
10. 4
11. perpendicular
12. parallel

Systematic Review 5D

1.
$$\begin{array}{r} 45 \\ +62 \\ \hline 107 \end{array}$$

2.
$$\begin{array}{r} 1 \\ 17 \\ +34 \\ \hline 51 \end{array}$$

3.
$$\begin{array}{r} 1 \\ 55 \\ +55 \\ \hline 110 \end{array}$$

4.
$$\begin{array}{r} 1 \\ 29 \\ +71 \\ \hline 100 \end{array}$$

5. $9 \times \underline{4} = 36$
6. $9 \times \underline{6} = 54$
7. $9 \times \underline{9} = 81$
8. $9 \times \underline{7} = 63$
9. $60 \div 10 = \underline{6}$
10. $15 \div 3 = \underline{5}$
11. $\dfrac{14}{2} = \underline{7}$
12. $\dfrac{30}{10} = \underline{3}$
13. $6 \div 3 = \underline{2}$
14. $45 \div 5 = \underline{9}$
15. $18 \div 2 = \underline{9}$
16. $35 \div 5 = \underline{7}$
17. yes
18. yes
19. $20 \div 5 = 4$ bags
20. $25 + 38 = 63$ problems

Systematic Review 5E

1.
$$\begin{array}{r} \overset{1}{1}9 \\ +88 \\ \hline 107 \end{array}$$

2.
$$\begin{array}{r} 54 \\ +13 \\ \hline 67 \end{array}$$

3.
$$\begin{array}{r} \overset{1}{4}3 \\ +67 \\ \hline 110 \end{array}$$

4.
$$\begin{array}{r} \overset{1}{7}4 \\ +48 \\ \hline 122 \end{array}$$

5. $9 \times \underline{8} = 72$
6. $9 \times \underline{10} = 90$
7. $9 \times \underline{2} = 18$
8. $9 \times \underline{10} = 90$
9. $50 \div 10 = \underline{5}$
10. $15 \div 5 = \underline{3}$
11. $\dfrac{18}{3} = \underline{6}$
12. $\dfrac{16}{2} = \underline{8}$
13. $30 \div 10 = \underline{3}$
14. $25 \div 5 = \underline{5}$
15. $27 \div 3 = \underline{9}$
16. $12 \div 2 = \underline{6}$
17. $\parallel; \perp$
18. no
19. $20 \div 2 = 10$ gumdrops
20. $9 \times 9 = 81$ sq ft

Systematic Review 5F

1.
$$\begin{array}{r} 24 \\ +35 \\ \hline 59 \end{array}$$

2.
$$\begin{array}{r} \overset{1}{1}3 \\ +19 \\ \hline 32 \end{array}$$

3.
$$\begin{array}{r} \overset{1}{8}1 \\ +79 \\ \hline 160 \end{array}$$

4.
$$\begin{array}{r} 65 \\ +42 \\ \hline 107 \end{array}$$

5. $9 \times \underline{2} = 18$
6. $9 \times \underline{6} = 54$
7. $9 \times \underline{8} = 72$
8. $9 \times \underline{1} = 9$
9. $40 \div 5 = \underline{8}$
10. $70 \div 10 = \underline{7}$
11. $\dfrac{21}{3} = \underline{7}$
12. $\dfrac{30}{3} = \underline{10}$
13. $100 \div 10 = \underline{10}$
14. $20 \div 5 = \underline{4}$
15. $24 \div 3 = \underline{8}$
16. $8 \div 2 = \underline{4}$
17. parallel, $\parallel$; perpendicular, $\perp$
18. no
19. $27 \div 3 = 9$ yd
20. $16 + 14 = 30$ children

Lesson Practice 6A

1. $9,18,27,36; \underline{4}$
2. $9,18,27; \underline{3}$
3. $9,18; \underline{2}$
4. $9,18,27,36,45,54; \underline{6}$
5. $81 \div 9 = \underline{9}$
6. $18 \div 9 = \underline{2}$
7. $63 \div 9 = \underline{7}$
8. $45 \div 9 = \underline{5}$
9. $90 \div 9 = \underline{10}$
10. $27 \div 9 = \underline{3}$
11. $9 \div 9 = \underline{1}$
12. $54 \div 9 = \underline{6}$
13. $72 \div 9 = \underline{8}$

14. $\frac{36}{9} = \underline{4}$

15. $\frac{81}{9} = \underline{9}$

16. $\frac{45}{9} = \underline{5}$

17. $90 \div 9 = \$10$ per day

18. $36 \div 9 = 4$ pieces

Lesson Practice 6B

1. 9,18,27,36,45;$\underline{5}$
2. 9,18;$\underline{2}$
3. 9,18,27;$\underline{3}$
4. 9,18,27,36,45,54,63,72;$\underline{8}$
5. $27 \div 9 = \underline{3}$
6. $54 \div 9 = \underline{6}$
7. $36 \div 9 = \underline{4}$
8. $72 \div 9 = \underline{8}$
9. $45 \div 9 = \underline{5}$
10. $9 \div 9 = \underline{1}$
11. $18 \div 9 = \underline{2}$
12. $81 \div 9 = \underline{9}$
13. $63 \div 9 = \underline{7}$
14. $\frac{90}{9} = \underline{10}$
15. $\frac{54}{9} = \underline{6}$
16. $\frac{36}{9} = \underline{4}$
17. $81 \div 9 = 9$ stories
18. $27 \div 9 = 3$ cakes

Lesson Practice 6C

1. 9,18,27,36,45,54,63,72,81,90;$\underline{10}$
2. 9,18,27,36,45,54,63,72,81;$\underline{9}$
3. 9;$\underline{1}$
4. 9,18,27,36,45,54,63;$\underline{7}$
5. $54 \div 9 = \underline{6}$
6. $9 \div 9 = \underline{1}$
7. $72 \div 9 = \underline{8}$

8. $45 \div 9 = \underline{5}$
9. $27 \div 9 = \underline{3}$
10. $36 \div 9 = \underline{4}$
11. $18 \div 9 = \underline{2}$
12. $63 \div 9 = \underline{7}$
13. $45 \div 9 = \underline{5}$
14. $\frac{72}{9} = \underline{8}$
15. $\frac{27}{9} = \underline{3}$
16. $\frac{81}{9} = \underline{9}$
17. $18 \div 2 = 9$ years
18. $54 \div 9 = 6$ teams

Systematic Review 6D

1. $54 \div 9 = \underline{6}$
2. $45 \div 9 = \underline{5}$
3. $63 \div 9 = \underline{7}$
4. $24 \div 3 = \underline{8}$
5. $35 \div 5 = \underline{7}$
6. $60 \div 10 = \underline{6}$
7. $36 \div 9 = \underline{4}$
8. $18 \div 2 = \underline{9}$
9. $72 \div 9 = \underline{8}$
10. $90 \div 9 = \underline{10}$
11. $\frac{25}{5} = \underline{5}$
12. $\frac{21}{3} = \underline{7}$
13. $5 \times \underline{1} = 5$
14. $9 \times \underline{9} = 81$
15. $3 \times \underline{4} = 12$
16. $10 \times \underline{10} = 100$
17. done
18.
$$\begin{array}{r} {}^6\!\!\!\not{7}\,{}^1 4 \\ -\ 3\ 8 \\ \hline 3\ 6 \end{array}$$
19.
$$\begin{array}{r} 59 \\ -41 \\ \hline 18 \end{array}$$

20.
$$\begin{array}{r} 67 \\ -25 \\ \hline 42 \end{array}$$

21. $81 \div 9 = 9$ wreaths

22. $95¢ - 87¢ = 8¢$ change

Systematic Review 6E

1. $36 \div 9 = \underline{4}$
2. $54 \div 9 = \underline{6}$
3. $18 \div 9 = \underline{2}$
4. $9 \div 9 = \underline{1}$
5. $15 \div 3 = \underline{5}$
6. $20 \div 5 = \underline{4}$
7. $40 \div 10 = \underline{4}$
8. $12 \div 2 = \underline{6}$
9. $27 \div 9 = \underline{3}$
10. $81 \div 9 = \underline{9}$
11. $\dfrac{72}{9} = \underline{8}$
12. $\dfrac{12}{3} = \underline{4}$
13. $2 \times \underline{9} = 18$
14. $10 \times \underline{5} = 50$
15. $3 \times \underline{7} = 21$
16. $5 \times \underline{9} = 45$
17.
$$\begin{array}{r} {}^{3}\cancel{4}\,{}^{1}3 \\ -19 \\ \hline 24 \end{array}$$
18.
$$\begin{array}{r} {}^{6}\cancel{7}\,{}^{1}8 \\ -59 \\ \hline 19 \end{array}$$
19.
$$\begin{array}{r} {}^{1} \\ 26 \\ +75 \\ \hline 101 \end{array}$$
20.
$$\begin{array}{r} {}^{5}\cancel{6}\,{}^{1}7 \\ -38 \\ \hline 29 \end{array}$$
21. $25 + 25 = 50$ pints
    $50 \div 10 = 5$ pints each

22. $14 \div 2 = 7$ qt
23. $27 \div 3 = 9$ yd
24. $32 - 14 = 18$ oranges

Systematic Review 6F

1. $90 \div 9 = \underline{10}$
2. $18 \div 9 = \underline{2}$
3. $36 \div 9 = \underline{4}$
4. $54 \div 9 = \underline{6}$
5. $27 \div 3 = \underline{9}$
6. $40 \div 5 = \underline{8}$
7. $72 \div 9 = \underline{8}$
8. $45 \div 5 = \underline{9}$
9. $80 \div 10 = \underline{8}$
10. $16 \div 2 = \underline{8}$
11. $\dfrac{63}{9} = \underline{7}$
12. $\dfrac{45}{9} = \underline{5}$
13. $9 \times \underline{8} = 72$
14. $3 \times \underline{8} = 24$
15. $5 \times \underline{10} = 50$
16. $9 \times \underline{0} = 0$
17.
$$\begin{array}{r} {}^{1} \\ 82 \\ +18 \\ \hline 100 \end{array}$$
18.
$$\begin{array}{r} {}^{2}\cancel{3}\,{}^{1}7 \\ -28 \\ \hline 9 \end{array}$$
19.
$$\begin{array}{r} {}^{5}\cancel{6}\,{}^{1}6 \\ -39 \\ \hline 27 \end{array}$$
20.
$$\begin{array}{r} 75 \\ +24 \\ \hline 99 \end{array}$$
21. $8 + 4 = 12$ pt
    $12 \div 2 = 6$ qt
22. $9 \times 3 = 27$ ft
23. $45 \div 9 = 5$ vans
24. $28¢ + 35¢ = 63¢$
    $63¢ \div 9¢ = 7$ treats

Lesson Practice 7A

1. done
2. $4 \times 6 = 24$ sq ft
3. $1 \times 2 = 2$ sq ft
4. $5 \times 5 = 25$ sq in
5. $3 \times 7 = 21$ sq in
6. $4 \times 10 = 40$ sq ft
7. $7 \times 5 = 35$ sq in
8. $9 \times 5 = 45$ sq yd

Lesson Practice 7B

1. $5 \times 10 = 50$ sq ft
2. $3 \times 9 = 27$ sq in
3. $9 \times 4 = 36$ sq in
4. $6 \times 10 = 60$ sq ft
5. $6 \times 7 = 42$ sq ft
6. $2 \times 10 = 20$ sq in
7. $10 \times 10 = 100$ sq in
8. $8 \times 5 = 40$ sq in

Lesson Practice 7C

1. $9 \times 7 = 63$ sq ft
2. $4 \times 5 = 20$ sq in
3. $10 \times 8 = 80$ sq in
4. $5 \times 6 = 30$ sq ft
5. $4 \times 7 = 28$ sq ft
6. $9 \times 5 = 45$ sq in
7. $8 \times 9 = 72$ sq yd
8. $10 \times 9 = 90$ sq yd

Systematic Review 7D

1. $9 \times 6 = 54$ sq ft
2. $6 \times 8 = 48$ sq in
3. $10 \times 5 = 50$ sq in
4. $6 \times 6 = 36$ sq mi
5. $36 \div 9 = \underline{4}$
6. $45 \div 5 = \underline{9}$

7. $18 \div 2 = \underline{9}$
8. $63 \div 9 = \underline{7}$
9. $27 \div 3 = \underline{9}$
10. $35 \div 5 = \underline{7}$
11. $\dfrac{18}{3} = \underline{6}$
12. $\dfrac{50}{10} = \underline{5}$
13. $6 \times \underline{7} = 42$
14. $6 \times \underline{5} = 30$
15. $4 \times \underline{8} = 32$
16. $6 \times \underline{2} = 12$
17.
$$\begin{array}{r} \overset{1}{3}8 \\ +26 \\ \hline 64 \end{array}$$
18.
$$\begin{array}{r} 15 \\ +84 \\ \hline 99 \end{array}$$
19.
$$\begin{array}{r} {}^{3}\!\!\not 4\,{}^{1}5 \\ -\ 1\ 6 \\ \hline 2\ 9 \end{array}$$
20.
$$\begin{array}{r} {}^{6}\!\not 7\,{}^{1}\!\not 1 \\ -\ 5\ 6 \\ \hline 1\ 5 \end{array}$$
21. $6 \times 5 = 30$ sq in
    $4 \times 4 = 16$ sq in
    $30 > 16$
    parallelogram
22. $43 - 28 = 15$ barrettes

Systematic Review 7E

1. $3 \times 2 = 6$ sq in
2. $7 \times 9 = 63$ sq ft
3. $8 \times 4 = 32$ sq mi
4. $10 \times 10 = 100$ sq ft
5. $12 \div 3 = \underline{4}$
6. $54 \div 9 = \underline{6}$
7. $24 \div 3 = \underline{8}$
8. $25 \div 5 = \underline{5}$
9. $14 \div 2 = \underline{7}$

10. $8 \div 1 = \underline{8}$

11. $\dfrac{81}{9} = \underline{9}$

12. $\dfrac{21}{3} = \underline{7}$

13. $4 \times \underline{6} = 24$

14. $6 \times \underline{10} = 60$

15. $6 \times \underline{7} = 42$

16. $4 \times \underline{7} = 28$

17.
```
   71
 +62
 133
```

18.
```
  ³4 ¹3
 - 2 5
   1 8
```

19.
```
   92
 +11
 103
```

20.
```
  ¹
   57
 +46
 103
```

21. parallel

22. $5 \times 3 = 15$ sq yd

Systematic Review 7F

1. $6 \times 7 = 42$ sq ft
2. $3 \times 8 = 24$ sq in
3. $10 \times 9 = 90$ sq ft
4. $3 \times 3 = 9$ sq mi
5. $27 \div 9 = \underline{3}$
6. $15 \div 3 = \underline{5}$
7. $30 \div 5 = \underline{6}$
8. $16 \div 2 = \underline{8}$
9. $72 \div 9 = \underline{8}$
10. $90 \div 10 = \underline{9}$
11. $\dfrac{20}{2} = \underline{10}$
12. $\dfrac{45}{9} = \underline{5}$

13. $4 \times \underline{8} = 32$

14. $6 \times \underline{8} = 48$

15. $6 \times \underline{6} = 36$

16. $4 \times \underline{4} = 16$

17.
```
  ¹2 ¹1
 -   9
   1 2
```

18.
```
    ¹
   76
 +54
 130
```

19.
```
   33
 +45
   78
```

20.
```
  ⁵6 ¹4
 - 2 5
   3 9
```

21. $14 \div 2 = 7$ qt

22. $30 - 16 = 14$ books

Lesson Practice 8A

1. $6,12,18;\underline{3}$
2. $6,12,18,24,30,36,42,48,54;\underline{9}$
3. $6,12;\underline{2}$
4. $6,12,18,24,30,36,42,48,54,60;\underline{10}$
5. $12 \div 6 = \underline{2}$
6. $6 \div 6 = \underline{1}$
7. $24 \div 6 = \underline{4}$
8. $36 \div 6 = \underline{6}$
9. $42 \div 6 = \underline{7}$
10. $18 \div 6 = \underline{3}$
11. $60 \div 6 = \underline{10}$
12. $24 \div 6 = \underline{4}$
13. $42 \div 6 = \underline{7}$
14. $\dfrac{54}{6} = \underline{9}$
15. $\dfrac{30}{6} = \underline{5}$
16. $\dfrac{48}{6} = \underline{8}$

17. $24 \div 6 = 4$ ants
18. $\$30 \div 6 = \$5$ a day

Lesson Practice 8B

1. 6,12,18,24,30;$\underline{5}$
2. 6;$\underline{1}$
3. 6,12,18,24;$\underline{4}$
4. 6,12,18,24,30,36,42,48;$\underline{8}$
5. $36 \div 6 = \underline{6}$
6. $60 \div 6 = \underline{10}$
7. $30 \div 6 = \underline{5}$
8. $18 \div 6 = \underline{3}$
9. $54 \div 6 = \underline{9}$
10. $42 \div 6 = \underline{7}$
11. $6 \div 6 = \underline{1}$
12. $24 \div 6 = \underline{4}$
13. $18 \div 6 = \underline{3}$
14. $\dfrac{30}{6} = \underline{5}$
15. $\dfrac{48}{6} = \underline{8}$
16. $\dfrac{12}{6} = \underline{2}$
17. $60 \div 6 = 10$ songs
18. $\$54 \div 6 = \$9$ each hour

Lesson Practice 8C

1. 6,12,18,24,30,36,42,48,54;$\underline{9}$
2. 6,12,18,24,30,36;$\underline{6}$
3. 6,12,18,24,30,36,42,48,54,60;$\underline{10}$
4. 6,12,18,24,30,36,42;$\underline{7}$
5. $18 \div 6 = \underline{3}$
6. $54 \div 6 = \underline{9}$
7. $6 \div 6 = \underline{1}$
8. $30 \div 6 = \underline{5}$
9. $12 \div 6 = \underline{2}$
10. $24 \div 6 = \underline{4}$

11. $42 \div 6 = \underline{7}$
12. $36 \div 6 = \underline{6}$
13. $48 \div 6 = \underline{8}$
14. $\dfrac{60}{6} = \underline{10}$
15. $\dfrac{54}{6} = \underline{9}$
16. $\dfrac{12}{6} = \underline{2}$
17. $\$48 \div 6 = \$8$ per friend
18. $18 \div 6 = 3$ ft
$3 \div 3 = 1$ yd

Systematic Review 8D

1. $18 \div 6 = \underline{3}$
2. $42 \div 6 = \underline{7}$
3. $54 \div 6 = \underline{9}$
4. $24 \div 3 = \underline{8}$
5. $25 \div 5 = \underline{5}$
6. $18 \div 2 = \underline{9}$
7. $54 \div 9 = \underline{6}$
8. $60 \div 10 = \underline{6}$
9. $48 \div 6 = \underline{8}$
10. $72 \div 9 = \underline{8}$
11. $\dfrac{21}{3} = \underline{7}$
12. $\dfrac{35}{5} = \underline{7}$
13. $12 \times 6 = 72$ sq ft
14. $7 \times 3 = 21$ sq in
15. $4 \times 4 = 16$ sq in
16.
```
     23        20+3
   ×36       × 30+6
   ‾1‾1‾     ‾100‾‾10‾
   128       100+20+8
    69       600+90+
   ‾‾‾‾‾     ‾‾‾‾‾‾‾‾‾
   828       800+20+8
```

17.
$$
\begin{array}{r}
78 \\
\times 34 \\
\hline
1_2 \\
23 \\
282 \\
214 \\
\hline
2652
\end{array}
$$

$$
\begin{array}{r}
70+8 \\
\times 30+4 \\
\hline
100_{200} \quad 30 \\
200 \quad +80+2 \\
2000+ \ 100 \ +40 \\
\hline
2000+ \ 600 \ +50+2
\end{array}
$$

18.
$$
\begin{array}{r}
65 \\
\times 15 \\
\hline
2 \\
305 \\
65 \\
\hline
975
\end{array}
$$

$$
\begin{array}{r}
60+5 \\
\times 10+5 \\
\hline
20 \\
300+00+5 \\
600+50+ \\
\hline
900+70+5
\end{array}
$$

19. $12 \times 15 = 180$ baby mice

20. $61 - 45 = 16$ sq ft

21. $36 \div 6 = 6$ afghans

22. $\$39 + \$28 = \$67$

Systematic Review 8E

1. $12 \div 6 = \underline{2}$
2. $60 \div 6 = \underline{10}$
3. $42 \div 6 = \underline{7}$
4. $24 \div 6 = \underline{4}$
5. $27 \div 9 = \underline{3}$
6. $40 \div 5 = \underline{8}$
7. $20 \div 10 = \underline{2}$
8. $12 \div 3 = \underline{4}$
9. $15 \div 3 = \underline{5}$
10. $30 \div 6 = \underline{5}$
11. $\dfrac{6}{6} = \underline{1}$
12. $\dfrac{12}{2} = \underline{6}$
13.
$$
\begin{array}{r}
1 \\
13 \\
+19 \\
\hline
32
\end{array}
$$
14.
$$
\begin{array}{r}
1 \\
28 \\
+49 \\
\hline
77
\end{array}
$$

15.
$$
\begin{array}{r}
^6\!\!\not7 \ ^1\!2 \\
- \ 2 \ 6 \\
\hline
4 \ 6
\end{array}
$$

16.
$$
\begin{array}{r}
^3\!\!\not4 \ ^1\!7 \\
- \ 3 \ 8 \\
\hline
9
\end{array}
$$

17.
$$
\begin{array}{r}
45 \\
\times 22 \\
\hline
1 \\
180 \\
80 \\
\hline
990
\end{array}
$$

$$
\begin{array}{r}
40+5 \\
\times 20+2 \\
\hline
10 \\
100+80+0 \\
800+00+ \\
\hline
900+90+0
\end{array}
$$

18.
$$
\begin{array}{r}
16 \\
\times 14 \\
\hline
2 \\
144 \\
16 \\
\hline
224
\end{array}
$$

$$
\begin{array}{r}
10+6 \\
\times 10+4 \\
\hline
20 \\
100+40+4 \\
100 \quad 60+ \\
\hline
200+20+4
\end{array}
$$

19.
$$
\begin{array}{r}
39 \\
\times \ 5 \\
\hline
14 \\
55 \\
\hline
195
\end{array}
$$

$$
\begin{array}{r}
30+9 \\
\times \qquad 5 \\
\hline
100 \quad 40 \\
+50+5 \\
\hline
100+90+5
\end{array}
$$

20. $30 \div 3 = 10$ yd
$\$6 \times 10 = \$60$

21. $14 \times 18 = 252$ sq in

22. $46 + 28 = 74$ mi

Systematic Review 8F

1. $48 \div 6 = \underline{8}$
2. $18 \div 6 = \underline{3}$
3. $12 \div 6 = \underline{2}$
4. $36 \div 6 = \underline{6}$
5. $72 \div 9 = \underline{8}$
6. $54 \div 6 = \underline{9}$
7. $27 \div 3 = \underline{9}$
8. $45 \div 5 = \underline{9}$
9. $70 \div 10 = \underline{7}$
10. $16 \div 2 = \underline{8}$
11. $\dfrac{42}{6} = \underline{7}$

12. $\dfrac{60}{6} = \underline{10}$

13. $\begin{array}{r} {}^1 8\,5 \\ +\ 1\,8 \\ \hline 1\,0\,3 \end{array}$

14. $\begin{array}{r} {}^3\cancel{4}\ {}^1 7 \\ -\ 3\,8 \\ \hline 9 \end{array}$

15. $\begin{array}{r} {}^1 4\,9 \\ +\ 2\,1 \\ \hline 7\,0 \end{array}$

16. $\begin{array}{r} {}^5\cancel{6}\ {}^1 4 \\ -\ 2\,5 \\ \hline 3\,9 \end{array}$

17.
$$\begin{array}{r} 3\,3 \\ \times 2\,4 \\ \hline 1 \\ 1\,2\,2 \\ 6\,6 \\ \hline 7\,9\,2 \end{array} \qquad \begin{array}{r} 3\,0+3 \\ \times 2\,0+4 \\ \hline 1\,0 \\ 1\,0\,0+2\,0+2 \\ 6\,0\,0+6\,0+ \\ \hline 7\,0\,0+9\,0+2 \end{array}$$

18.
$$\begin{array}{r} 4\,4 \\ \times 1\,4 \\ \hline 1\ 1 \\ 1\,6\,6 \\ 4\,4 \\ \hline 6\,1\,6 \end{array} \qquad \begin{array}{r} 4\,0+4 \\ \times 1\,0+4 \\ \hline 1\,0\,0\quad 1\,0 \\ 1\,0\,0+6\,0+6 \\ 4\,0\,0+4\,0+ \\ \hline 6\,0\,0+1\,0+6 \end{array}$$

19.
$$\begin{array}{r} 1\,5 \\ \times 1\,5 \\ \hline 1\ 2 \\ 5\,5 \\ 1\,5 \\ \hline 2\,2\,5 \end{array} \qquad \begin{array}{r} 1\,0+5 \\ \times 1\,0+5 \\ \hline 1\,0\,0\quad 2\,0 \\ +5\,0+5 \\ 1\,0\,0+5\,0+ \\ \hline 2\,0\,0+2\,0+5 \end{array}$$

20. $24 \div 6 = 4$ turns

21. $\$35 \times 14 = \$490$

22. $42 \div 6 = 7$ ft

Lesson Practice 9A

1. done
2. $4 \times 4 = 16$
   $16 \div 2 = \underline{8}$ sq in
3. $2 \times 7 = 14$
   $14 \div 2 = \underline{7}$ sq mi

4. $3 \times 6 = 18$
   $18 \div 2 = \underline{9}$ sq ft
5. $4 \times 5 = 20$
   $20 \div 2 = \underline{10}$ sq in
6. $9 \times 2 = 18$
   $18 \div 2 = \underline{9}$ sq ft
7. $8 \times 2 = 16$
   $16 \div 2 = \underline{8}$ sq yd
8. $1 \times 2 = 2$
   $2 \div 2 = \underline{1}$ sq in
9. $2 \times 2 = 4$
   $4 \div 2 = \underline{2}$ sq mi
10. $2 \times 4 = 8$
    $8 \div 2 = \underline{4}$ sq in

Lesson Practice 9B

1. $3 \times 4 = 12$
   $12 \div 2 = \underline{6}$ sq ft
2. $2 \times 6 = 12$
   $12 \div 2 = \underline{6}$ sq in
3. $1 \times 8 = \underline{8}$
   $8 \div 2 = \underline{4}$ sq mi
4. $2 \times 10 = 20$
   $20 \div 2 = \underline{10}$ sq ft
5. $2 \times 5 = 10$
   $10 \div 2 = \underline{5}$ sq in
6. $10 \times 1 = \underline{10}$
   $10 \div 2 = \underline{5}$ sq in
7. $2 \times 3 = 6$
   $6 \div 2 = \underline{3}$ sq yd
8. $3 \times 6 = \underline{18}$
   $18 \div 2 = \underline{9}$ sq ft
9. $3 \div 3 = 1$ yd
   $6 \div 3 = 2$ yd
10. $2 \times 1 = 2$
    $2 \div 2 = \underline{1}$ sq yd

Lesson Practice 9C

1. $1 \times 4 = 4$
   $4 \div 2 = 2$ sq ft
2. $2 \times 5 = 10$
   $10 \div 2 = 5$ sq in
3. $2 \times 9 = 18$
   $18 \div 2 = 9$ sq mi
4. $4 \times 4 = 16$
   $16 \div 2 = 8$ sq ft
5. $2 \times 8 = 16$
   $16 \div 2 = 8$ sq in
6. $2 \times 4 = 8$
   $8 \div 2 = 4$ sq in
7. $2 \times 2 = 4$
   $4 \div 2 = 2$ sq yd
8. $2 \times 7 = 14$
   $14 \div 2 = 7$ sq ft
9. $4 \times 3 = 12$
   $12 \div 2 = 6$ sq ft
   so 6 plants
10. $6 \times 3 = 18$
    $18 \div 2 = 9$ sq yd

Systematic Review 9D

1. $1 \times 4 = 4$
   $4 \div 2 = 2$ sq ft
2. $2 \times 3 = 6$
   $6 \div 2 = 3$ sq in
3. $5 \times 7 = 35$ sq in
4. $36 \div 6 = \underline{6}$
5. $42 \div 6 = \underline{7}$
6. $18 \div 6 = \underline{3}$
7. $54 \div 6 = \underline{9}$
8. $63 \div 9 = \underline{7}$
9. $40 \div 5 = \underline{8}$
10. $\frac{27}{3} = \underline{9}$
11. $\frac{80}{10} = \underline{8}$

12. $\begin{array}{r} 50 \\ \times 32 \\ \hline 100 \\ 150 \\ \hline 1600 \end{array}$ $50 \times 32 = \underline{1600}$
13. $\begin{array}{r} 16 \\ \times 18 \\ \hline 4 \\ 88 \\ 16 \\ \hline 288 \end{array}$ $16 \times 18 = \underline{288}$
14. $\begin{array}{r} 28 \\ \times 22 \\ \hline 146 \\ 46 \\ \hline 616 \end{array}$ $28 \times 22 = \underline{616}$
15. $\begin{array}{r} 32 \\ \times 17 \\ \hline 214 \\ 32 \\ \hline 544 \end{array}$ $32 \times 17 = \underline{544}$
16. $\begin{array}{r} 23 \\ 26 \\ +37 \\ \hline 86 \end{array}$ $23+26+37 = 86$
17. $\begin{array}{r} 12 \\ 59 \\ +31 \\ \hline 102 \end{array}$ $12+59+31 = 102$
18. $\begin{array}{r} 15 \\ 15 \\ 44 \\ +24 \\ \hline 98 \end{array}$ $15+15+44+24 = 98$
19. $\begin{array}{r} 34 \\ 56 \\ 11 \\ + 9 \\ \hline 110 \end{array}$ $34+56+11+9 = 110$
20. $5+11+4+8 = 28$ pages

Systematic Review 9E
1. $6 \times 7 = 42$ sq ft
2. $2 \times 6 = 12$
   $12 \div 2 = 6$ sq in
3. $4 \times 8 = 32$ sq in
4. $30 \div 6 = \underline{5}$
5. $48 \div 6 = \underline{8}$
6. $12 \div 6 = \underline{2}$
7. $24 \div 6 = \underline{4}$
8. $72 \div 9 = \underline{8}$
9. $27 \div 9 = \underline{3}$
10. $\frac{35}{5} = \underline{7}$
11. $\frac{3}{3} = \underline{1}$
12. $\begin{array}{r} ^5\cancel{6}\,{}^1 0 \\ -\ 3\ 1 \\ \hline 2\ 9 \end{array}$ $60 - 31 = 29$
13. $\begin{array}{r} ^1\cancel{2}\,{}^1 7 \\ -\ 1\ 8 \\ \hline 9 \end{array}$ $27 - 18 = 9$
14. $\begin{array}{r} ^4\cancel{5}\,{}^1 2 \\ -\ 2\ 7 \\ \hline 2\ 5 \end{array}$ $52 - 27 = 25$
15. $\begin{array}{r} ^1 6 \\ 7\ 2 \\ 3\ 8 \\ +\ 3\ 1 \\ \hline 1\ 5\ 7 \end{array}$ $16 + 72 + 38 + 31 = 157$
16. $\begin{array}{r} ^1 8\ 0 \\ 1\ 4 \\ 6\ 8 \\ 4\ 3 \\ +\ 7\ 2 \\ \hline 2\ 7\ 7 \end{array}$ $80 + 14 + 68 + 43 + 72 = 277$
17. $\begin{array}{r} 5 \\ ^3 9 \\ 8\ 4 \\ 7\ 1 \\ +\ 2\ 6 \\ \hline 2\ 2\ 5 \end{array}$ $5 + 39 + 84 + 71 + 26 = 225$
18. $54 \div 9 = 6$ teams
19. $7 \times 2 = 14$ pints
20. 2 sets

Systematic Review 9F
1. $5 \times 5 = 25$ sq ft
2. $9 \times 10 = 90$ sq ft
3. $3 \times 4 = 12$
   $12 \div 2 = 6$ sq yd
4. $60 \div 6 = \underline{10}$
5. $36 \div 9 = \underline{4}$
6. $18 \div 3 = \underline{6}$
7. $\frac{20}{5} = \underline{4}$
8. $45 \times 16 = 720$
9. $52 \times 28 = 1,456$
10. $76 \times 54 = 4,104$
11. $33 + 75 + 44 + 67 = 219$
12. $83 + 90 + 45 + 25 + 17 = 260$
13. $26 + 43 + 31 + 57 + 14 = 171$
14. $\$45 \div \$5 = 9$
    $9 - 1$ (himself) $= 8$ friends
15. $21 \div 3 = 7$ boxes
16. $42 \div 6 = 7$ treats
17. $11 + 9 + 13 = 33$ points
18. $62 - 49 = 13$¢
19. $12 \times 25 = 300$ cookies
20. no

Lesson Practice 10A
1. 4,8,12,16,20;$\underline{5}$
2. 4,8;$\underline{2}$
3. 4,8,12,16,20,24,28;$\underline{7}$
4. done
5. $\begin{array}{r} 10 \\ 4\overline{)40} \\ \underline{-40} \\ 0 \end{array}$

6.
$$\begin{array}{r} 1 \\ 4\overline{)4} \\ -4 \\ \hline 0 \end{array}$$

7.
$$\begin{array}{r} 5 \\ 4\overline{)20} \\ -20 \\ \hline 0 \end{array}$$

8.
$$\begin{array}{r} 8 \\ 4\overline{)32} \\ -32 \\ \hline 0 \end{array}$$

9.
$$\begin{array}{r} 7 \\ 4\overline{)28} \\ -28 \\ \hline 0 \end{array}$$

10. $12 \div 4 = \underline{3}$

11. $36 \div 4 = \underline{9}$

12. $40 \div 4 = \underline{10}$

13. $\dfrac{16}{4} = \underline{4}$

14. $\dfrac{32}{4} = \underline{8}$

15. $\dfrac{36}{4} = \underline{9}$

16. $24 \div 4 = 6$ cups

17. $28 \div 4 = 7$ days

18. $16 \div 4 = 4$ chairs

6.
$$\begin{array}{r} 2 \\ 4\overline{)8} \\ -8 \\ \hline 0 \end{array}$$

7.
$$\begin{array}{r} 6 \\ 4\overline{)24} \\ -24 \\ \hline 0 \end{array}$$

8.
$$\begin{array}{r} 3 \\ 4\overline{)12} \\ -12 \\ \hline 0 \end{array}$$

9.
$$\begin{array}{r} 9 \\ 4\overline{)36} \\ -36 \\ \hline 0 \end{array}$$

10. $20 \div 4 = \underline{5}$

11. $28 \div 4 = \underline{7}$

12. $32 \div 4 = \underline{8}$

13. $\dfrac{40}{4} = \underline{10}$

14. $\dfrac{4}{4} = \underline{1}$

15. $\dfrac{24}{4} = \underline{6}$

16. $36 \div 4 = 9$ cards

17. $\$20 \div \$4 = 5$ hours

18. $40 \div 4 = 10$ horses

Lesson Practice 10B

1. $4,8,12,16,20,24,28,32,36;\underline{9}$

2. $4;\underline{1}$

3. $4,8,12;\underline{3}$

4.
$$\begin{array}{r} 8 \\ 4\overline{)32} \\ -32 \\ \hline 0 \end{array}$$

5.
$$\begin{array}{r} 4 \\ 4\overline{)16} \\ -16 \\ \hline 0 \end{array}$$

Lesson Practice 10C

1. $4,8,12,16,20,24;\underline{6}$

2. $4,8,12,16;\underline{4}$

3. $4,8,12,16,20,24,28,32;\underline{8}$

4.
$$\begin{array}{r} 2 \\ 4\overline{)8} \\ -8 \\ \hline 0 \end{array}$$

5.
$$\begin{array}{r} 7 \\ 4\overline{)28} \\ -28 \\ \hline 0 \end{array}$$

6.
$$4\overline{)40} \quad \begin{array}{r} 10 \\ -40 \\ \hline 0 \end{array}$$

7.
$$4\overline{)36} \quad \begin{array}{r} 9 \\ -36 \\ \hline 0 \end{array}$$

8.
$$4\overline{)20} \quad \begin{array}{r} 5 \\ -20 \\ \hline 0 \end{array}$$

9.
$$4\overline{)4} \quad \begin{array}{r} 1 \\ -4 \\ \hline 0 \end{array}$$

10. $12 \div 4 = \underline{3}$

11. $24 \div 4 = \underline{6}$

12. $16 \div 4 = \underline{4}$

13. $\dfrac{8}{4} = \underline{2}$

14. $\dfrac{28}{4} = \underline{7}$

15. $\dfrac{40}{4} = \underline{10}$

16. $32 \div 4 = 8$ cages

17. $28 \div 4 = 7$ chocolates

18. $8 \div 4 = 2$ years

Systematic Review 10D

1.
$$4\overline{)36} \quad \begin{array}{r} 9 \\ -36 \\ \hline 0 \end{array}$$

2.
$$4\overline{)20} \quad \begin{array}{r} 5 \\ -20 \\ \hline 0 \end{array}$$

3.
$$4\overline{)16} \quad \begin{array}{r} 4 \\ -16 \\ \hline 0 \end{array}$$

4.
$$4\overline{)28} \quad \begin{array}{r} 7 \\ -28 \\ \hline 0 \end{array}$$

5. $36 \div 6 = \underline{6}$

6. $14 \div 2 = \underline{7}$

7. $\dfrac{15}{3} = \underline{5}$

8. $\dfrac{81}{9} = \underline{9}$

9.
$$\begin{array}{r} {}^{1}3 \\ 25 \\ 37 \\ +42 \\ \hline 117 \end{array}$$

10.
$$\begin{array}{r} {}^{2}3\ {}^{1}1 \\ -2\ 2 \\ \hline 9 \end{array}$$

11.
$$\begin{array}{r} {}^{4}5\ {}^{1}8 \\ -3\ 9 \\ \hline 1\ 9 \end{array}$$

12. $45 \times 15 = 675$

13. $25 \times 12 = 300$ sq ft

14. $12 \times 6 = 72$ sq in

15. $2 \times 5 = 10$
$10 \div 2 = 5$ sq in

16. $8 \times 4 = 32$ quarters

17. $8 \div 4 = 2$ gallons

18. $6 \times 4 = 24$ quarts

Systematic Review10E

1.
$$4\overline{)40} \quad \begin{array}{r} 10 \\ -40 \\ \hline 0 \end{array}$$

2.
$$\begin{array}{r} 3 \\ 4\overline{)12} \\ -12 \\ \hline 0 \end{array}$$

3.
$$\begin{array}{r} 8 \\ 4\overline{)32} \\ -32 \\ \hline 0 \end{array}$$

4.
$$\begin{array}{r} 6 \\ 4\overline{)24} \\ -24 \\ \hline 0 \end{array}$$

5. $100 \div 10 = \underline{10}$

6. $35 \div 5 = \underline{7}$

7. $\dfrac{27}{9} = \underline{3}$

8. $\dfrac{54}{6} = \underline{9}$

9.
$$\begin{array}{r} {}^{2} \\ 46 \\ 14 \\ 23 \\ +17 \\ \hline 100 \end{array}$$

10.
$$\begin{array}{r} {}^{6}\!\!\not{7}\,{}^{1}6 \\ -\ 4\ 7 \\ \hline 2\ 9 \end{array}$$

11. $64 \times 32 = 2,048$

12. $43 \times 84 = 3,612$

13. $5 \times 4 = 20$
$20 \div 2 = 10$ sq ft

14. $23 \times 28 = 644$ sq mi

15. $3 \times 6 = 18$
$18 \div 2 = 9$ sq ft

16. $\$28 \div 4 = \$7$

17. $20 \div 4 = \$5$

18. $24 \times 12 = 288$ books

19. $45 \div 5 = 9$ shots

20. $6 + 8 + 10 + 12 = 36$ quarts
$36 \div 4 = 9$ gallons

Systematic Review10F

1.
$$\begin{array}{r} 4 \\ 44\overline{)16} \\ -16 \\ \hline 0 \end{array}$$

2.
$$\begin{array}{r} 1 \\ 44\overline{)4} \\ -4 \\ \hline 0 \end{array}$$

3.
$$\begin{array}{r} 10 \\ 44\overline{)40} \\ -40 \\ \hline 0 \end{array}$$

4.
$$\begin{array}{r} 2 \\ 44\overline{)8} \\ -8 \\ \hline 0 \end{array}$$

5. $18 \div 2 = \underline{9}$

6. $24 \div 3 = \underline{8}$

7. $\dfrac{42}{6} = \underline{7}$

8. $\dfrac{72}{9} = \underline{8}$

9.
$$\begin{array}{r} {}^{2}3\ 8 \\ 4\ 1 \\ 1\ 2 \\ +\ \ 9 \\ \hline 1\ 0\ 0 \end{array}$$

10.
$$\begin{array}{r} {}^{1}5\ 6 \\ 2\ 4 \\ 1\ 8 \\ +\ 2\ 1 \\ \hline 1\ 1\ 9 \end{array}$$

11.
$$\begin{array}{r} {}^{8}\!\not{9}\,{}^{1}1 \\ -\ 2\ 7 \\ \hline 6\ 4 \end{array}$$

12.
$$\begin{array}{r} 7\ 5 \\ -2\ 5 \\ \hline 5\ 0 \end{array}$$

13. $7 \times 2 = 14$
$14 \div 2 = 7$ sq ft

14. $16 \times 9 = 144$ sq mi

15. $2 \times 4 = 8$

$8 \div 2 = 4$ sq in

16. $12 \div 4 = 3$ packs

17. $32 \div 4 = \$8$

18. $46 \times 46 = 2,116$ sq ft

19. $16 + 19 + 10 + 15 = 60$

$60 \times 2 = 120$ minutes

20. $24 \div 4 = 6$ teams

Lesson Practice 11A

1. done

2. $5 + 4 + 6 = 15$

$15 \div 3 = \underline{5}$

3. $8 + 8 + 5 = 21$

$21 \div 3 = \underline{7}$

4. $9 + 11 = 20$

$20 \div 2 = \underline{10}$

5. $2 + 3 + 1 = 6$

$6 \div 3 = \underline{2}$

6. $10 + 6 = 16$

$16 \div 2 = \underline{8}$

7. $10 + 7 + 6 + 9 = 32$

$32 \div 4 = \underline{8}$

8. $2 + 8 + 5 + 9 + 1 = 25$

$25 \div 5 = \underline{5}$

9. $7 + 6 + 14 + 9 = 36$

$36 \div 4 = \underline{9}$

10. $9 + 15 + 6 = 30$

$30 \div 3 = 10$ cards per month

11. $4 + 5 + 3 + 4 + 4 = 20$

$20 \div 5 = 4$ emails per day

12. $8 + 6 + 5 + 10 + 13 + 12 = 54$

$54 \div 6 = 9$ pages per day

Lesson Practice 11B

1. $2 + 5 + 5 = 12$

$12 \div 3 = \underline{4}$

2. $10 + 5 + 9 = 24$

$24 \div 3 = \underline{8}$

3. $11 + 6 + 10 = 27$

$27 \div 3 = \underline{9}$

4. $2 + 8 = 10$

$10 \div 2 = \underline{5}$

5. $7 + 9 + 14 + 6 = 36$

$36 \div 4 = \underline{9}$

6. $7 + 5 = 12$

$12 \div 2 = \underline{6}$

7. $10 + 3 + 6 + 1 = 20$

$20 \div 4 = \underline{5}$

8. $8 + 5 + 8 + 7 = 28$

$28 \div 4 = \underline{7}$

9. $3 + 3 + 6 + 9 + 9 = 30$

$30 \div 5 = \underline{6}$

10. $10 + 11 + 9 + 10 = 40$

$40 \div 4 = 10$ jewels per mine

11. $8 + 7 + 10 + 7 = 32$

$32 \div 4 = 8$ parts per vehicle

12. $8 + 7 + 4 + 10 + 9 + 10 = 48$

$48 \div 6 = 8$ in per month

Lesson Practice 11C

1. $8 + 10 + 12 = 30$

$30 \div 3 = \underline{10}$

2. $7 + 6 + 8 = 21$

$21 \div 3 = \underline{7}$

3. $1 + 2 + 6 = 9$

$9 \div 3 = \underline{3}$

4. $1 + 5 = 6$

$6 \div 2 = \underline{3}$

5. $1 + 3 + 4 + 8 = 16$

$16 \div 4 = \underline{4}$

6. $6 + 8 = 14$

$14 \div 2 = \underline{7}$

7. $1+3+6+7+13 = 30$
   $30 \div 5 = \underline{6}$

8. $4+2+6+8 = 20$
   $20 \div 4 = \underline{5}$

9. $9+12+2+5 = 28$
   $28 \div 4 = \underline{7}$

10. $9+12+8+11 = 40$
    $40 \div 4 = 10$ points per quarter

11. $10+9+12+5+9 = 45$
    $45 \div 5 = $ 9 tickets

12. $2+3+4+6+8+1 = 24$
    $24 \div 6 = $ 4 in

13.
    $$\begin{array}{r} {}^{6}\cancel{7}\,{}^{1}2 \\ -\ 3\ 4 \\ \hline 3\ 8 \end{array}$$

14. $15 \times 47 = 705$

15. $28 \times 18 = 504$

16. $7 \times 8 = 56$ sq ft

17. $2 \times 8 = 16$
    $16 \div 2 = 8$ sq in

18. $3+2+3+5+2 = 15$
    $15 \div 5 = 3$ hours per day

19. $24 \div 4 = 6$ gallons

20. perpendicular

Systematic Review 11D

1. $2+1+3+6 = 12$
   $12 \div 4 = \underline{3}$

2. $5+7+3+10+15 = 40$
   $40 \div 5 = \underline{8}$

3. $9+4+5 = 18$
   $18 \div 3 = \underline{6}$

4. $7 \times \underline{7} = 49$

5. $8 \times \underline{6} = 48$

6. $7 \times \underline{8} = 56$

7.
   $$\begin{array}{r} 4 \\ 9\overline{)36} \\ -36 \\ \hline 0 \end{array}$$

8.
   $$\begin{array}{r} 10 \\ 5\overline{)50} \\ -50 \\ \hline 0 \end{array}$$

9.
   $$\begin{array}{r} 7 \\ 6\overline{)42} \\ -42 \\ \hline 0 \end{array}$$

10. $18 \div 2 = \underline{9}$

11. $63 \div 9 = \underline{7}$

12. $\dfrac{30}{6} = \underline{5}$

Systematic Review 11E

1. $9+6+7+2 = 24$
   $24 \div 4 = \underline{6}$

2. $1+3+9+8+7+8 = 36$
   $36 \div 6 = \underline{6}$

3. $8+6+13 = 27$
   $27 \div 3 = \underline{9}$

4. $8 \times \underline{8} = 64$

5. $7 \times \underline{9} = 63$

6. $8 \times \underline{5} = 40$

7.
   $$\begin{array}{r} 3 \\ 9\overline{)27} \\ -27 \\ \hline 0 \end{array}$$

8.
   $$\begin{array}{r} 8 \\ 9\overline{)72} \\ -72 \\ \hline 0 \end{array}$$

9.
   $$\begin{array}{r} 2 \\ 2\overline{)4} \\ -4 \\ \hline 0 \end{array}$$

10. $45 \div 9 = \underline{5}$

11. $18 \div 6 = \underline{3}$

12. $\dfrac{70}{10} = \underline{7}$

13. $37 \times 17 = 629$

14.
$$\begin{array}{r} 48 \\ -21 \\ \hline 27 \end{array}$$

15.
$$\begin{array}{r} {}^{4}\!\!\!/5\,{}^{1}\!3 \\ -29 \\ \hline 24 \end{array}$$

16. $10 \times 12 = 120$ sq ft

17. $4 \times 3 = 12$
$12 \div 2 = 6$ sq ft

18. $36 \div 4 = \$9$

19. $29 + 25 = 54$
$54 \div 9 = 6$ stickers each

20. $8 \times \underline{7} = 56$

13. $55 \times 27 = 1,485$

14. $62 \times 38 = 2,356$

15.
$$\begin{array}{r} {}^{8}\!9\,{}^{1}\!5 \\ -46 \\ \hline 49 \end{array}$$

16. $8 \times 8 = 64$ sq in

17. $1 \times 6 = 6$
$6 \div 2 = 3$ sq mi

18. $30 \div 3 = 10$ yd

19. $9 \times 3 = 27$
$27 - 5 = 22$ ft

20. E,F,H,M,N,Z
(possibly I, depending on style)

Systematic Review 11F

1. $3 + 4 + 5 + 4 = 16$
$16 \div 4 = \underline{4}$

2. $1 + 1 + 2 + 3 + 3 = 10$
$10 \div 5 = \underline{2}$

3. $4 + 7 + 4 = 15$
$15 \div 3 = \underline{5}$

4. $8 \times \underline{9} = 72$

5. $7 \times \underline{8} = 56$

6. $7 \times \underline{7} = 49$

7.
$$\begin{array}{r} 2 \\ 9\overline{)18} \\ -18 \\ \hline 0 \end{array}$$

8.
$$\begin{array}{r} 2 \\ 6\overline{)12} \\ -12 \\ \hline 0 \end{array}$$

9.
$$\begin{array}{r} 9 \\ 6\overline{)54} \\ -54 \\ \hline 0 \end{array}$$

10. $25 \div 5 = \underline{5}$

11. $80 \div 10 = \underline{8}$

12. $\dfrac{81}{9} = \underline{9}$

Lesson Practice 12A

1. 8,16,24;<u>3</u>

2. 7,14,21,28,35,42,49;<u>7</u>

3. 8,16;<u>2</u>

4.
$$\begin{array}{r} 8 \\ 7\overline{)56} \\ -56 \\ \hline 0 \end{array}$$

5.
$$\begin{array}{r} 8 \\ 8\overline{)64} \\ -64 \\ \hline 0 \end{array}$$

6.
$$\begin{array}{r} 6 \\ 7\overline{)42} \\ -42 \\ \hline 0 \end{array}$$

7.
$$\begin{array}{r} 5 \\ 8\overline{)40} \\ -40 \\ \hline 0 \end{array}$$

8.
$$\begin{array}{r} 4 \\ 8\overline{)32} \\ -32 \\ \hline 0 \end{array}$$

9.
$$
\begin{array}{r}
9 \\
7\overline{)63} \\
-63 \\
\hline
0
\end{array}
$$

10. $56 \div 8 = \underline{7}$

11. $35 \div 7 = \underline{5}$

12. $48 \div 8 = \underline{6}$

13. $\dfrac{28}{7} = \underline{4}$

14. $\dfrac{21}{7} = \underline{3}$

15. $\dfrac{72}{8} = \underline{9}$

16. $56 \div 8 = 7$ weeks

17. $14 \div 7 = 2$ cookies

18. $48 \div 8 = 6$ octopuses

Lesson Practice 12B

1. $7,14,21,28,35,42,49,56,63;\underline{9}$

2. $8,16,24,32,40,48,56,64;\underline{8}$

3. $7,14,21,28,35,42;\underline{6}$

4.
$$
\begin{array}{r}
10 \\
7\overline{)70} \\
-70 \\
\hline
0
\end{array}
$$

5.
$$
\begin{array}{r}
3 \\
8\overline{)24} \\
-24 \\
\hline
0
\end{array}
$$

6.
$$
\begin{array}{r}
7 \\
7\overline{)49} \\
-49 \\
\hline
0
\end{array}
$$

7.
$$
\begin{array}{r}
6 \\
8\overline{)48} \\
-48 \\
\hline
0
\end{array}
$$

8.
$$
\begin{array}{r}
1 \\
8\overline{)8} \\
-8 \\
\hline
0
\end{array}
$$

9.
$$
\begin{array}{r}
8 \\
7\overline{)56} \\
-56 \\
\hline
0
\end{array}
$$

10. $40 \div 8 = \underline{5}$

11. $28 \div 7 = \underline{4}$

12. $72 \div 8 = \underline{9}$

13. $\dfrac{21}{7} = \underline{3}$

14. $\dfrac{35}{7} = \underline{5}$

15. $\dfrac{56}{8} = \underline{7}$

16. $32 \div 8 = 4$ hours

17. $56 \div 7 = 8$ weeks

18. $63 \div 7 = 9$ in

Lesson Practice 12C

1. $8,16,24,32,40,48,56;\underline{7}$

2. $7,14,21,28,35,42,49,56;\underline{8}$

3. $8,16,24,32,40,48,56,64,72;\underline{9}$

4.
$$
\begin{array}{r}
9 \\
7\overline{)63} \\
-63 \\
\hline
0
\end{array}
$$

5.
$$
\begin{array}{r}
4 \\
8\overline{)32} \\
-32 \\
\hline
0
\end{array}
$$

6.
$$
\begin{array}{r}
3 \\
7\overline{)21} \\
-21 \\
\hline
0
\end{array}
$$

7.
$$
\begin{array}{r}
8 \\
8\overline{)64} \\
-64 \\
\hline
0
\end{array}
$$

8.
$$\begin{array}{r} 3 \\ \overline{)24} \\ -24 \\ \hline 0 \end{array}$$

9.
$$\begin{array}{r} 7 \\ 7\overline{)49} \\ -49 \\ \hline 0 \end{array}$$

10. $16 \div 8 = \underline{2}$

11. $42 \div 7 = \underline{6}$

12. $40 \div 8 = \underline{5}$

13. $\dfrac{35}{7} = \underline{5}$

14. $\dfrac{14}{7} = \underline{2}$

15. $\dfrac{80}{8} = \underline{10}$

16. $70 \div 7 = 10$ troops

17. $48 \div 8 = 6$ words

18. $28 \div 7 = 4$ miles

Systematic Review 12D

1.
$$\begin{array}{r} 8 \\ 7\overline{)56} \\ -56 \\ \hline 0 \end{array}$$

2.
$$\begin{array}{r} 8 \\ 8\overline{)64} \\ -64 \\ \hline 0 \end{array}$$

3.
$$\begin{array}{r} 7 \\ 8\overline{)56} \\ -56 \\ \hline 0 \end{array}$$

4.
$$\begin{array}{r} 7 \\ 7\overline{)49} \\ -49 \\ \hline 0 \end{array}$$

5. $16 \div 4 = \underline{4}$

6. $36 \div 6 = \underline{6}$

7. $\dfrac{72}{9} = \underline{8}$

8. $\dfrac{21}{7} = \underline{3}$

9. $7 + 1 + 2 + 6 + 4 = 20$
$20 \div 5 = \underline{4}$

10. $10 + 13 + 9 + 9 + 8 + 11 = 60$
$60 \div 6 = 10$

11. $6 + 5 + 13 = 24$
$24 \div 3 = 8$

12.
$$\begin{array}{r} 1\overset{1}{2}4 \\ +369 \\ \hline 493 \end{array}$$

13.
$$\begin{array}{r} \overset{1}{7}\overset{1}{8}1 \\ +319 \\ \hline 1100 \end{array}$$

14.
$$\begin{array}{r} 3\overset{1}{3}5 \\ +126 \\ \hline 461 \end{array}$$

15.
$$\begin{array}{r} 4\overset{1}{0}4 \\ +278 \\ \hline 682 \end{array}$$

16. $8 \times 16 = 128$ oz

17. $36 \div 4 = \$9$

18. $12 \div 4 = 3$ hours

19. $9 \times 9 = 81$ sq in

20. $18 \times 3 = 54$ ft

Systematic Review 12E

1.
$$\begin{array}{r} 4 \\ 7\overline{)28} \\ -28 \\ \hline 0 \end{array}$$

2.
$$\begin{array}{r} 6 \\ 8\overline{)48} \\ -48 \\ \hline 0 \end{array}$$

3.
$$\begin{array}{r} 9 \\ 7\overline{)63} \\ -63 \\ \hline 0 \end{array}$$

4.
$$\begin{array}{r} 6 \\ 7\overline{)42} \\ -42 \\ \hline 0 \end{array}$$

5. $35 \div 5 = \underline{7}$

6. $27 \div 3 = \underline{9}$

7. $\dfrac{54}{9} = \underline{6}$

8. $\dfrac{56}{7} = \underline{8}$

9. $9 + 4 + 5 + 6 = 24$
$24 \div 4 = \underline{6}$

10. $1 + 2 + 3 + 4 + 4 + 5 + 6 + 7 = 32$
$32 \div 8 = \underline{4}$

11. $4 + 5 + 9 = 18$
$18 \div 3 = \underline{6}$

12.
$$\begin{array}{r} 4\,{}^1\!2\,{}^1\!1 \\ -2\ 08 \\ \hline 2\ 13 \end{array}$$

13.
$$\begin{array}{r} 6\,{}^3\!4\,{}^1\!2 \\ -1\ 27 \\ \hline 5\ 15 \end{array}$$

14.
$$\begin{array}{r} {}^6\!7\,{}^1\!89 \\ -\ 3\ 94 \\ \hline 3\ 95 \end{array}$$

15.
$$\begin{array}{r} {}^2\!3\,{}^1\!03 \\ -\ 163 \\ \hline 1\ 40 \end{array}$$

16. $2 \times 9 = 18$
$18 \div 2 = 9$ sq ft

17. $25 \times 16 = 400$ oz

18. $20 \div 4 = 5$ mistakes

19. $251 + 317 = 568$
$568 - 179 = 389$ nuts

20. $49 \div 7 = 7$ books

Systematic Review 12F

1.
$$\begin{array}{r} 5 \\ 7\overline{)35} \\ -35 \\ \hline 0 \end{array}$$

2.
$$\begin{array}{r} 9 \\ 8\overline{)72} \\ -72 \\ \hline 0 \end{array}$$

3.
$$\begin{array}{r} 2 \\ 7\overline{)14} \\ -14 \\ \hline 0 \end{array}$$

4.
$$\begin{array}{r} 3 \\ 7\overline{)21} \\ -21 \\ \hline 0 \end{array}$$

5. $40 \div 8 = \underline{5}$

6. $16 \div 8 = \underline{2}$

7. $\dfrac{45}{5} = \underline{9}$

8. $\dfrac{64}{8} = \underline{8}$

9. $2 + 4 + 6 + 8 + 10 = 30$
$30 \div 5 = 6$

10. $1 + 1 + 2 + 2 + 4 + 4 + 5 + 5 = 24$
$24 \div 8 = \underline{3}$

11. $4 + 10 + 16 = 30$
$30 \div 3 = \underline{10}$

12. $17 \times 25 = \underline{425}$

13. $48 \times 36 = \underline{1,728}$

14. $89 \times 43 = \underline{3,827}$

15. $78 \times 87 = \underline{6,786}$

16. $5 \times 9 = 45$ sq in

17. 1 lb $= 16$ oz
20 oz > 16 oz

18. $18 \div 2 = 9$ jars

19. 32 qt $\div 4 = 8$ gal
8 gal $- 2$ gal $= 6$ gal

20. $12 \times 24 = 288$ pencils

Lesson Practice 13A

1. done
2. $9+11=20$
   $20 \div 2 = 10$
   $10 \times 6 = 60$ sq in
3. $6+10=16$
   $16 \div 2 = 8$
   $8 \times 4 = 32$ sq ft
4. $6+12=18$
   $18 \div 2 = 9$
   $9 \times 7 = 63$ sq ft
5. $3+5=8$
   $8 \div 2 = 4$
   $4 \times 2 = 8$ sq ft
6. $8+10=18$
   $18 \div 2 = 9$
   $9 \times 5 = 45$ sq in
7. $7+9=16$
   $16 \div 2 = 8$
   $8 \times 5 = 40$ sq in
8. $1+3=4$
   $4 \div 2 = 2$
   $2 \times 2 = 4$ sq mi

Lesson Practice 13B

1. $2+4=6$
   $6 \div 2 = 3$
   $3 \times 3 = 9$ sq in
2. $3+17=20$
   $20 \div 2 = 10$
   $10 \times 6 = 60$ sq ft
3. $5+9=14$
   $14 \div 2 = 7$
   $7 \times 5 = 35$ sq ft
4. $4+10=14$
   $14 \div 2 = 7$
   $7 \times 3 = 21$ sq in
5. $8+12=20$
   $20 \div 2 = 10$
   $10 \times 11 = 110$ sq ft

6. $7+11=18$
   $18 \div 2 = 9$
   $9 \times 4 = 36$ sq in
7. $2+6=8$
   $8 \div 2 = 4$
   $4 \times 6 = 24$ sq ft
   24 plants
8. $5+7=12$
   $12 \div 2 = 6$
   $6 \times 6 = 36$ sq in

Lesson Practice 13C

1. $4+10=14$
   $14 \div 2 = 7$
   $7 \times 4 = 28$ sq in
2. $1+11=12$
   $12 \div 2 = 6$
   $6 \times 3 = 18$ sq ft
3. $6+8=14$
   $14 \div 2 = 7$
   $7 \times 5 = 35$ sq ft
4. $4+6=10$
   $10 \div 2 = 5$
   $5 \times 4 = 20$ sq in
5. $5+11=16$
   $16 \div 2 = 8$
   $8 \times 10 = 80$ sq ft
6. $2+14=16$
   $16 \div 2 = 8$
   $8 \times 6 = 48$ sq ft
7. $1+5=6$
   $6 \div 2 = 3$
   $3 \times 2 = 6$ sq mi
8. $6+14=20$
   $20 \div 2 = 10$
   $10 \times 15 = 150$ sq ft
   150 tiles

Systematic Review 13D

1. $5 + 7 = 12$
   $12 \div 2 = 6$
   $6 \times 6 = 36$ sq in
2. $3 + 15 = 18$
   $18 \div 2 = 9$
   $9 \times 5 = 45$ sq ft
3. $27 \div 9 = \underline{3}$
4. $36 \div 6 = \underline{6}$
5. $28 \div 7 = \underline{4}$
6. $45 \div 5 = \underline{9}$
7. $56 \div 8 = \underline{7}$
8. $49 \div 7 = \underline{7}$
9. $\frac{16}{2} = \underline{8}$
10. $\frac{42}{6} = \underline{7}$
11. $6 + 5 + 7 + 10 = 28$
    $28 \div 4 = \underline{7}$
12. $2 + 6 + 1 + 4 + 3 + 2 = 18$
    $18 \div 6 = \underline{3}$
13. $1 + 5 + 15 = 21$
    $21 \div 3 = \underline{7}$
14. $7 \times 20 = \underline{140}$
15. $13 \times 20 = \underline{260}$
16. $9 \times 30 = \underline{270}$
17. $24 \div 4 = 6$ horses
18. $60 \div 6 = 10$ songs
19. $20 \times 12 = 240$ months
20. $\$90 - \$65 = \$25$

Systematic Review 13E

1. $3 + 11 = 14$
   $14 \div 2 = 7$
   $7 \times 7 = 49$ sq ft
2. $25 \times 18 = 450$ sq in
3. $63 \div 9 = \underline{7}$
4. $72 \div 9 = \underline{8}$
5. $2 \div 2 = \underline{1}$
6. $18 \div 3 = \underline{6}$
7. $32 \div 4 = \underline{8}$

8. $25 \div 5 = \underline{5}$
9. $\frac{48}{6} = \underline{8}$
10. $\frac{56}{7} = \underline{8}$
11. $\begin{array}{r} {}^{1}78 \\ +\ 3\ 34 \\ \hline 5\ 12 \end{array}$
12. $\begin{array}{r} 5\,{}^{1}\!2\,{}^{1}\!1 \\ -\ 4\ \ 13 \\ \hline 1\ 08 \end{array}$
13. $\begin{array}{r} 7\,{}^{1}\!08 \\ +\ 2\ 22 \\ \hline 9\ 30 \end{array}$
14. $9 \times 20 = \underline{180}$
15. $34 \times 20 = \underline{680}$
16. $11 \times 30 = \underline{330}$
17. $63 \div 7 = 9$ days
18. $24 \div 4 = 6$ ft
19. $11 + 15 + 4 + 10 + 5 = 45$ dogs
    $45 \div 5 = 9$ dogs
20. $13 \times 16 = 208$ oz

Systematic Review 13F

1. $2 + 18 = 20$
   $20 \div 2 = 10$
   $10 \times 6 = 60$ sq ft
2. $2 \times 3 = 6$
   $6 \div 2 = 3$ sq ft
3. $35 \div 7 = \underline{5}$
4. $36 \div 9 = \underline{4}$
5. $54 \div 6 = \underline{9}$
6. $64 \div 8 = \underline{8}$
7. $48 \div 8 = \underline{6}$
8. $24 \div 6 = \underline{4}$
9. $\frac{50}{5} = \underline{10}$
10. $\frac{21}{7} = \underline{3}$
11. $16 \div 4 = \underline{4}$
12. $36 \div 4 = \underline{9}$

13. $24 \div 3 = \underline{8}$
14. $6 \times 30 = \underline{180}$
15. $21 \times 20 = \underline{420}$
16. $42 \times 30 = \underline{1,260}$
17. $32 \div 8 = 4$ cakes
18. $20 \div 4 = \$5$
19. $50 - 20 = 30$ quarters
    $6 \times 4 = 24$ quarters
    He needs enough for 30, but
    has only enough for 24.
20. $419 + 495 = 914$ words

Lesson Practice 14A

1. done
2. 35,201
3. 765,892
4. 4,265,143
5. done
6. $10,000 + 100 + 20 + 9$
7. $100,000 + 90,000 + 5,000 + 300 + 20 + 8$
8. $1,000,000 + 700,000 + 80,000 + 6,000 + 200 + 1$
9. done
10. 28,616
11. 4,300,400
12. 6,815,231

Lesson Practice 14B

1. 2,794
2. 16,302
3. 651,741
4. 2,540,000
5. $7,000 + 800 + 1$
6. $40,000 + 1,000 + 400 + 50 + 6$
7. $200,000 + 30,000 + 8,000 + 100 + 90 + 9$
8. $5,000,000 + 300,000 + 60,000 + 5,000$
9. 3,021
10. 45,615

11. 5,400,000
12. 8,131,528

Lesson Practice 14C

1. 1,224
2. 43,638
3. 247,000
4. 3,122,472
5. $3,000 + 200 + 50 + 6$
6. $50,000 + 600 + 4$
7. $700,000 + 50,000 + 4,000 + 700 + 50 + 3$
8. $2,000,000 + 100,000 + 10,000 + 7,000 + 200 + 40 + 9$
9. 1,838
10. 33,230
11. 2,350,000
12. 4,652,893

Systematic Review 14D

1. 1,652
2. $6,000,000 + 300,000 + 40,000 + 100 + 20 + 9$
3. $4 + 12 = 16$
   $16 \div 2 = 8$
   $8 \times 9 = 72$ sq ft
4. $2 \times 6 = 12$
   $12 \div 2 = 6$ sq in
5. $42 \div 7 = \underline{6}$
6. $54 \div 6 = \underline{9}$
7. $81 \div 9 = \underline{9}$
8. $40 \div 8 = \underline{5}$
9. $14 \div 7 = \underline{2}$
10. $12 \div 4 = \underline{3}$
11. $\dfrac{15}{3} = \underline{5}$
12. $\dfrac{70}{7} = \underline{10}$
13. done
14. $17 \times 100 = 1,700$

15. $8 \times 200 = 1,600$
16. $478 - 182 = 296$ marbles
17. $555 + 555 = 1,110$ ft
18. $24 \div 4 = 6$ cookies

Systematic Review 14E

1. 25,611
2. $1,000,000 + 100,000 + 70,000 + 4,000$
3. $2 + 4 = 6$
   $6 \div 2 = 3$
   $3 \times 3 = 9$ sq in
4. $12 \times 12 = 144$ sq mi
5. $35 \div 5 = \underline{7}$
6. $18 \div 6 = \underline{3}$
7. $28 \div 7 = \underline{4}$
8. $27 \div 9 = \underline{3}$
9. $72 \div 8 = \underline{9}$
10. $20 \div 5 = \underline{4}$
11. $\dfrac{8}{4} = \underline{2}$
12. $\dfrac{12}{3} = \underline{4}$
13. $23 \times 100 = \underline{2,300}$
14. $14 \times 200 = \underline{2,800}$
15. $9 \times 200 = \underline{1,800}$
16. $8 \times 11 = 88$ sq ft
17. $12 \times 10 = 120$ logs
18. $14 - 6 = 8$ qt
    $8 \times 2 = 16$ pt

Systematic Review 14F

1. 14,715
2. $4,000,000 + 700,000 + 10,000 + 1,000 + 300 + 40$
3. $9 + 11 = 20$
   $20 \div 2 = 10$
   $10 \times 10 = 100$ sq in
4. $23 \times 38 = 874$ sq ft
5. $90 \div 10 = \underline{9}$

6. $18 \div 9 = \underline{2}$
7. $49 \div 7 = \underline{7}$
8. $42 \div 6 = \underline{7}$
9. $14 \div 2 = \underline{7}$
10. $7 \div 1 = \underline{7}$
11. $\dfrac{36}{6} = \underline{6}$
12. $\dfrac{28}{4} = \underline{7}$
13. $7 \times 200 = \underline{1,400}$
14. $33 \times 200 = \underline{6,600}$
15. $15 \times 100 = \underline{1,500}$
16. parallel
17. $15 \times 16 = 240$ oz
18. $7 + 15 + 13 + 9 + 6 = 50$
    $50 \div 5 = 10$ points

Lesson Practice 15A

1. done
2. 11,368; eleven thousand, three hundred sixty-eight
3. 3,000,000; three million
4. 6,000,000,000; six billion
5. done
6. $3 \times 10,000 + 1 \times 100 + 8 \times 1$
7. $2 \times 1,000,000,000 + 5 \times 100,000,000 + 4 \times 10,000,000 + 2 \times 1,000,000$
8. $6 \times 1,000,000,000,000,000$
9. done
10. 2,113,000
11. 7,945,000,000
12. 2,000,000,000,112

Lesson Practice 15B

1. 50,940; fifty thousand, nine hundred forty
2. 672,800; six hundred seventy-two thousand eight hundred
3. 94,000,000; ninety-four million

4. 2,648,000,000,000; two trillion, six hundred forty-eight billion
5. $5 \times 100,000,000,000 + 1 \times 10 + 9 \times 1$
6. $1 \times 1,000,000,000,000 + 7 \times 100,000,000,000 + 8 \times 10,000,000,000 + 3 \times 1,000,000,000$
7. $7 \times 10,000,000,000 + 2 \times 1,000,000,000 + 3 \times 100,000,000 + 5 \times 10,000,000$
8. $7 \times 1,000,000,000,000 + 6 \times 100 + 2 \times 10$
9. 59,140
10. 1,000,089
11. 32,000,477,000
12. 7,891,000,000,000

Lesson Practice 15C

1. 60,000,006; sixty million, six
2. 495,006,200; four hundred ninety-five million, six thousand, two hundred
3. 36,000,000; thirty-six million
4. 3,003,000,000,000; three trillion, three billion
5. $7 \times 100,000,000,000 + 2 \times 10,000,000,000 + 5 \times 1,000,000,000 + 7 \times 10,000,000 + 8 \times 1,000,000$
6. $4 \times 10,000,000,000,000 + 3 \times 100,000,000,000 + 1 \times 10,000,000,000 + 6 \times 1,000,000,000$
7. $1 \times 1,000,000,000 + 4 \times 100,000,000 + 6 \times 10,000,000 + 5 \times 1,000,000 + 9 \times 100,000$
8. $3 \times 100,000,000,000,000 + 7 \times 10,000,000,000,000 + 1 \times 1,000,000,000,000 + 8 \times 100,000,000,000$
9. 256,094
10. 6,000,851,000,000
11. 874,000,320
12. 1,067,000,000

Systematic Review 15D

1. 547,000,000,000; five hundred forty-seven billion
2. $5 \times 100,000,000,000,000 + 6 \times 10,000,000,000,000 + 4 \times 1,000,000,000,000$
3. $21 \div 7 = \underline{3}$
4. $30 \div 6 = \underline{5}$
5. $56 \div 8 = \underline{7}$
6. $45 \div 9 = \underline{5}$
7. done
8. $\times$
9. $+$
10. $-$
11.
$$\begin{array}{r} {}^{1,}{}^{1}582 \\ + 3,624 \\ \hline 5,206 \end{array}$$
12.
$$\begin{array}{r} 7,132 \\ +5,333 \\ \hline 12,465 \end{array}$$
13.
$$\begin{array}{r} {}^{1}2,{}^{1}852 \\ + 4,263 \\ \hline 7,115 \end{array}$$
14. $22 \times 40 = 880$
15. $11 \times 60 = 660$
16. $12 \times 40 = 480$
17. $200 - 17 = 183$ pieces
18. $1,436 + 1,529 = 2965$ mi

Systematic Review 15E

1. 107,873; one hundred seven thousand eight hundred seventy-three
2. $3 \times 1,000,000,000,000,000 + 5 \times 100$
3. $56 \div 7 = \underline{8}$
4. $48 \div 6 = \underline{8}$
5. $30 \div 5 = \underline{6}$
6. $21 \div 3 = \underline{7}$
7. $\times$
8. $-$
9. $\div$

10. +

11.
$$\begin{array}{r} 1,{}^3\!4\,{}^1\!2\,6 \\ -\quad 8\,7\,3 \\ \hline 5\,5\,3 \end{array}$$

12.
$$\begin{array}{r} {}^3\!4,{}^1\!2{}^7\!8\,{}^1\!3 \\ -\qquad 9\,5\,5 \\ \hline 3,\,3\;2\;8 \end{array}$$

13.
$$\begin{array}{r} {}^5\!6,{}^1\!2{}^1\!3{}^3\!4\,{}^1\!1 \\ -\qquad\;\; 3\;7\;8 \\ \hline 5,\;8\quad 6\;3 \end{array}$$

14. $14 \times 20 = 280$
15. $20 \times 100 = 2,000$
16. $45 \times 100 = 4,500$
17. $27 + 38 + 34 = 99$
    $99 \div 3 = 33$ runs
18. $36 \times \$6 = \$216$
    $36 \div 4 = 9$ gal

Systematic Review 15F

1. 8,472,600,000; eight billion,
   four hundred seventy-two million
   six hundred thousand
2. $2 \times 1,000,000,000,000 +$
   $2 \times 10,000,000,000 + 7 \times 1,000,000,000$
3. $63 \div 9 = \underline{7}$
4. $24 \div 8 = \underline{3}$
5. $42 \div 6 = \underline{7}$
6. $49 \div 7 = \underline{7}$
7. $\div$
8. $\div$
9. $+$
10. $-$
11.
$$\begin{array}{r} {}^1\!6,{}^1\!7\,3\,2 \\ 3,\,1\,5\,2 \\ +\,7,\,3\,2\,1 \\ \hline 1\,7,\,2\,0\,5 \end{array}$$

12.
$$\begin{array}{r} 5,9\,8\,9 \\ -\quad 6\,3\,2 \\ \hline 5,3\,5\,7 \end{array}$$

13.
$$\begin{array}{r} {}^1\!5,{}^1\!2\,3\,2 \\ 7,1\,1\,1 \\ +\,3,7\,6\,5 \\ \hline 1\,6,1\,0\,8 \end{array}$$

14. $11 \times 60 = 660$
15. $21 \times 40 = 840$
16. $40 \times 200 = 8,000$
17. $8 + 12 = 20$
    $20 \div 2 = 10$
    $10 \times 25 = 250$ sq ft
18. $2 \times 16 = 32$ ounces lost
    32 quarters $\div 4 = \$8$

Lesson Practice 16A

1. done
2.
$$\begin{array}{r} 5\text{r. }3 \\ 4\,\overline{)23} \\ \underline{20} \\ 3 \end{array}$$

3.
$$\begin{array}{r} 8\text{r. }3 \\ 7\,\overline{)59} \\ \underline{56} \\ 3 \end{array}$$

4.
$$\begin{array}{r} 1\text{r. }6 \\ 7\,\overline{)13} \\ \underline{7} \\ 6 \end{array}$$

5.
$$\begin{array}{r} 4\text{r. }1 \\ 8\,\overline{)33} \\ \underline{32} \\ 1 \end{array}$$

6.
$$\begin{array}{r} 2\text{r. }5 \\ 8\,\overline{)21} \\ \underline{16} \\ 5 \end{array}$$

7.
$$
\begin{array}{r}
9\text{r. }4 \\
9\overline{)85} \\
\underline{81} \\
4
\end{array}
$$

8.
$$
\begin{array}{r}
2\text{r. }2 \\
9\overline{)20} \\
\underline{18} \\
2
\end{array}
$$

9.
$$
\begin{array}{r}
7\text{r. }1 \\
9\overline{)64} \\
\underline{63} \\
1
\end{array}
$$

10.
$$
\begin{array}{r}
6\text{r. }3 \\
6\overline{)39} \\
\underline{36} \\
3
\end{array}
$$

11.
$$
\begin{array}{r}
6\text{r. }4 \\
6\overline{)40} \\
\underline{36} \\
4
\end{array}
$$

12.
$$
\begin{array}{r}
2\text{r. }1 \\
6\overline{)13} \\
\underline{12} \\
1
\end{array}
$$

13.
$$
\begin{array}{r}
9\text{r. }1 \\
6\overline{)46} \\
\underline{45} \\
1
\end{array}
$$

14.
$$
\begin{array}{r}
3\text{r. }4 \\
5\overline{)19} \\
\underline{15} \\
4
\end{array}
$$

15.
$$
\begin{array}{r}
9\text{r. }3 \\
5\overline{)48} \\
\underline{45} \\
3
\end{array}
$$

16. 9r. 3 ; 9 cookies each, 3 left over
$$
\begin{array}{r}
8\overline{)75} \\
\underline{72} \\
3
\end{array}
$$

17. 6r. 1; 6 days, 1 left over
$$
\begin{array}{r}
2\overline{)13} \\
\underline{12} \\
1
\end{array}
$$

18. 9r. 3 ; 9 dollars each, 3 bills left over
$$
\begin{array}{r}
5\overline{)48} \\
\underline{45} \\
3
\end{array}
$$

Lesson Practice 16B

1.
$$
\begin{array}{r}
2\text{r. }1 \\
3\overline{)7} \\
\underline{6} \\
1
\end{array}
$$

2.
$$
\begin{array}{r}
4\text{r. }2 \\
3\overline{)14} \\
\underline{12} \\
2
\end{array}
$$

3.
$$
\begin{array}{r}
9\text{r. }2 \\
3\overline{)29} \\
\underline{27} \\
2
\end{array}
$$

4.
$$
\begin{array}{r}
4\text{r. }1 \\
4\overline{)17} \\
\underline{16} \\
1
\end{array}
$$

5.
$$
\begin{array}{r}
4\text{r. }1 \\
2\overline{)9} \\
\underline{8} \\
1
\end{array}
$$

6.
$$
\begin{array}{r}
6\text{r. }1 \\
6\overline{)37} \\
\underline{36} \\
1
\end{array}
$$

7.
$$
\begin{array}{r}
4\text{r. }1 \\
7\overline{)29} \\
\underline{28} \\
1
\end{array}
$$

8.
```
    2r. 3
 7)17
    14
     3
```

9.
```
   10r. 1
 5)51
   50
    1
```

10.
```
    1r. 2
 5)7
    5
    2
```

11.
```
    8r. 2
 8)66
   64
    2
```

12.
```
    3r. 3
 8)27
   24
    3
```

13.
```
    8r. 3
 9)75
   72
    3
```

14.
```
    1r. 5
 9)14
    9
    5
```

15.
```
    3r. 1
 3)10
    9
    1
```

16. 5r. 1; 5 cookies each, 1 left over
```
 4)21
   20
    1
```

17. 4r. 2; no, 2 extra
```
 6)26
   24
    2
```

18. 4r. 3; 4 teams, 3 extra
```
 9)39
   36
    3
```

Lesson Practice 16C

1.
```
    9r. 1
 2)19
   18
    1
```

2.
```
    1r. 1
 2)3
    2
    1
```

3.
```
    3r. 1
 5)16
   15
    1
```

4.
```
    7r. 2
 5)37
   35
    2
```

5.
```
    8r. 4
 9)76
   72
    4
```

6.
```
    3r. 2
 9)29
   27
    2
```

7.
```
    9r. 2
 4)38
   36
    2
```

8.
```
    3r.1
 4)13
   12
    1
```

9.
```
    1r. 2
  7|9
    7
    2
```

10.
```
    3r. 1
  7|22
    21
     1
```

11.
```
    5r. 1
  3|16
    15
     1
```

12.
```
    6r. 2
  3|20
    18
     2
```

13.
```
    2r. 3
  8|19
    16
     3
```

14.
```
    9r. 7
  8|79
    72
     7
```

15.
```
    8r. 2
  6|50
    48
     2
```

16.
```
    7r. 2 ; 7 weeks, 2 days
  7|51
    49
     2
```

17.
```
    6r. 2 ; $6 apiece, $2 left over
  5|32
    30
     2
```

18.
```
    3r. 2 ; 4 cars
  5|17
    15
     2
```

Systematic Review 16D

1.
```
    7r. 2
  3|23
    21
     2
```

2.
```
    10r. 1
  3|31
    30
     1
```

3.
```
    3r. 2
  6|20
    18
     2
```

4.
```
    5r. 1
  2|11
    10
     1
```

5.
```
    7r. 2
  4|30
    28
     2
```

6.
```
    5r. 5
  7|40
    35
     5
```

7.
```
    6r. 3
  9|57
    54
     3
```

8.
```
    2r. 2
  5|12
    10
     2
```

9.
```
    5r. 3
  8|43
    40
     3
```

10.
```
     1  1
    6,718
   +2,452
    9,170
```

11.
$$5,{}^2\cancel{3}{}^9\cancel{7}{}^1\cancel{0}\;{}^1 2$$
$$-1,\;2\;\;3\;\;8$$
$$\overline{4,\;0\;\;6\;\;4}$$

12.
$$4,{}^5\cancel{6}{}^{10}\cancel{1}\;{}^1 2$$
$$-3,\;5\;\;2\;\;6$$
$$\overline{1,\;0\;\;8\;\;6}$$

13. $11 \times 700 = 7,700$

14. $12 \times 300 = 3,600$

15. $20 \times 500 = 10,000$

16. $6,217,000,000$

17.
$$\begin{array}{r} 9\text{r. }1 \\ 2\overline{)19} \\ \underline{18} \\ 1 \end{array}$$

9 full loads, 1 bucket on last trip

18. $1,227 + 2,341 = 3,568$ animals

Systematic Review 16E

1.
$$\begin{array}{r} 2\text{r. }2 \\ 4\overline{)10} \\ \underline{8} \\ 2 \end{array}$$

2.
$$\begin{array}{r} 6\text{r. }3 \\ 7\overline{)45} \\ \underline{42} \\ 3 \end{array}$$

3.
$$\begin{array}{r} 5\text{r. }4 \\ 9\overline{)49} \\ \underline{45} \\ 4 \end{array}$$

4.
$$\begin{array}{r} 7\text{r. }4 \\ 6\overline{)46} \\ \underline{42} \\ 4 \end{array}$$

5.
$$\begin{array}{r} 8\text{r. }1 \\ 3\overline{)25} \\ \underline{24} \\ 1 \end{array}$$

6.
$$\begin{array}{r} 6\text{r. }2 \\ 8\overline{)50} \\ \underline{48} \\ 2 \end{array}$$

7.
$$\begin{array}{r} 5\text{r. }2 \\ 5\overline{)27} \\ \underline{25} \\ 2 \end{array}$$

8.
$$\begin{array}{r} 2\text{r. }1 \\ 2\overline{)5} \\ \underline{4} \\ 1 \end{array}$$

9.
$$\begin{array}{r} 5\text{r. }5 \\ 6\overline{)35} \\ \underline{30} \\ 5 \end{array}$$

10.
$$\begin{array}{r} {}^1 2,482 \\ +7,902 \\ \hline 10,384 \end{array}$$

11.
$$\begin{array}{r} 7,{}^1 059 \\ +3,470 \\ \hline 10,529 \end{array}$$

12.
$$\begin{array}{r} 6,997 \\ -2,961 \\ \hline 4,036 \end{array}$$

13. $11 \times 600 = 6,600$

14. $13 \times 200 = 2,600$

15. $22 \times 300 = 6,600$

16. $3 \times 1,000,000,000,000 + 4 \times 100,000,000,000 +$
$9 \times 10,000,000,000 + 1 \times 1,000,000,000$

17.
$$\begin{array}{r} 3\text{r. }1;\;\; 3\text{ CDs, \$1 left over} \\ 8\overline{)25} \\ \underline{24} \\ 1 \end{array}$$

18. $20 + 13 + 8 + 7 = 48$
$48 \div 4 = 12$ lb

Systematic Review 16F

1.
$$\begin{array}{r} 3\text{r. }1 \\ 2\overline{)7} \\ \underline{6} \\ 1 \end{array}$$

2.
$$\begin{array}{r} 4\text{r. }4 \\ 5\overline{)24} \\ \underline{20} \\ 4 \end{array}$$

3.
$$\begin{array}{r} 7\text{r. }6 \\ 8\overline{)62} \\ \underline{56} \\ 6 \end{array}$$

4.
$$\begin{array}{r} 9\text{r. }1 \\ 3\overline{)28} \\ \underline{27} \\ 1 \end{array}$$

5.
$$\begin{array}{r} 9\text{r. }7 \\ 9\overline{)88} \\ \underline{81} \\ 7 \end{array}$$

6.
$$\begin{array}{r} 8\text{r. }1 \\ 4\overline{)33} \\ \underline{32} \\ 1 \end{array}$$

7.
$$\begin{array}{r} 9\text{r. }2 \\ 7\overline{)65} \\ \underline{63} \\ 2 \end{array}$$

8.
$$\begin{array}{r} 1\text{r. }2 \\ 6\overline{)8} \\ \underline{6} \\ 2 \end{array}$$

9.
$$\begin{array}{r} 7\text{r. }4 \\ 7\overline{)53} \\ \underline{49} \\ 4 \end{array}$$

10.
$$\begin{array}{r} 4,5{,}^{10}4^{1}42 \\ -3{,}\,9\,7\,1 \\ \hline 1{,}\,1\,7\,1 \end{array}$$

11.
$$\begin{array}{r} 9{,}0^{5}6\,{}^{1}5 \\ -4{,}0\,1\,8 \\ \hline 5{,}0\,4\,7 \end{array}$$

12.
$$\begin{array}{r} {}^{1}8{,}932 \\ +6{,}823 \\ \hline 15{,}755 \end{array}$$

13. $10 \times 700 = 7{,}000$

14. $21 \times 300 = 6{,}300$

15. $22 \times 400 = 8{,}800$

16. $7{,}349{,}000$

17.
$$\begin{array}{r} 8\text{r. }2\,; \ 8 \text{ presents, 2 bows left over} \\ 3\overline{)26} \\ \underline{24} \\ 2 \end{array}$$

18. rectangle:
$8 \times 10 = 80$ sq ft
trapezoid:
$5 + 7 = 12$
$12 \div 2 = 6$
$6 \times 6 = 36$ sq ft
$36 + 80 = 116$ sq ft total
$116 < 300$, so yes

Lesson Practice 17A

1. done

2. done

3.
$$\begin{array}{rr} 33 & 30+3 \\ \times\ 3 & \times\quad 3 \\ \hline 99 & 90+9 \end{array}$$

4.
$$\begin{array}{rr} 3 & 3 \\ \times 33 & \times 30+3 \\ \hline 9 & +9 \\ \underline{90} & \underline{90} \\ 99 & 90+9 = 99 \end{array}$$

5.
$$\begin{array}{rr} 431 & 431 \\ \times\ \ 2 & \times\qquad\quad 2 \\ \hline 862 & 800+60+2 \end{array}$$

6.
```
      2        2
  ×431    ×400+30+1
     2         2
    60        60
   800       800
   862    800+60+2 = 862
```

7. done

8. done

9.
```
     20
  5)100
   -100
      0
```

10.
```
     30
  3)90
   -90
     0
```

11.
```
     60
  6)360
   -360
      0
```

12.
```
     40
  4)160
   -160
      0
```

Lesson Practice 17B

1.
```
    32      30+2
  × 3      ×   3
   96      90+6
```

2.
```
     3        3
  ×32     ×30+2
    6        6
   90       90
   96    90+6 = 96
```

3.
```
   438     400+30+ 8
  × 2     ×       2
   876     800+60+16
```

4.
```
      2        2
  ×438    ×400+30+8
    16        16
    60        60
   800       800
   876    800+70+6 = 876
```

5.
```
    13
   129     100+20+ 9
  ×  4    ×       4
   516     400+80+36
```

6.
```
     4         4
  ×129    ×100+20+ 9
    36        3 6
    80        8 0
   400       40 0
   516    400+80+36 = 516
```

7.
```
     20
  3)60
   -60
     0
```

8.
```
     90
  9)810
   -810
      0
```

9.
```
     30
  8)240
   -240
      0
```

10.
```
     40
  2)80
   -80
     0
```

11.
```
     60
  5)300
   -300
      0
```

12.
```
     30
  7)210
   -210
      0
```

Lesson Practice 19C

1.
```
      5              3
     00            ×105
   100r. 2          15
  3⟌317             00
    300            300
     17            315 +2 = 317
     15
      2
```

2.
```
      7              7
     10            ×117
   100r. 5           49
  7⟌824              70
    700             700
    124             819 +5 = 824
     70
     54
     49
      5
```

3.
```
      1              4
     10            ×111
   100r. 1           4
  4⟌445             40
    400            400
     45            444 +1 = 445
     40
      5
      4
      1
```

4.
```
      6              4
    60r. 2          ×66
  4⟌266             24
    240            240
     26            264 +2 = 266
     24
      2
```

5.
```
      0              6
     10            ×110
   100r. 95          0
  6⟌665             60
    600            600
     65            660 +5 = 665
     60
      5
      0
      5
```

6.
```
      1              9
     90            ×91
  9⟌819             9
    810            810
      9            819
      9
      0
```

7. $802 \div 2 = 401$ jugs

8. $550 \div 5 = 110$ cages

Systematic Review19D

1.
```
      8
     00
   100r. 3
  4⟌435
    400
     35
     32
      3
```

2.
```
      4
    ×108
     32
     00
    400
    432 +3 = 435
```

3.
```
        4
      40r. 2
  6 )266
      240
       26
       24
        2
```

4.
```
        6
     ×44
      24
     240
     264 +2 = 266
```

5.
```
      1 1 1
    2,962
   +3,148
    6,110
```

6.
```
    1,¹250
   +  351
    1,601
```

7.
```
    ³,⁰76
   +1,942
    5,018
```

8. $6 \times 2,000 = 12,000$

9. $2 \times 5,280 = 10,560$

10. $8 \times 5,280 = 42,240$

11. parallel

12. $6 \times 1,000,000,000,000 + 5 \times 100,000,000,000 + 8 \times 10,000,000,000 + 3 \times 1,000,000,000$

13. 6,000,000,705

14. $7,821 - 4,568 = 3,253$ miles

15. $69 \div 6 = 11$ r. 3; 11 teams, 3 left

16. $3 \times 2,000 = 6,000$ lb

Systematic Review19E

1.
```
       20
      300
   3 )960
      900
       60
       60
        0
```

2.
```
        3
    ×320
        0
       60
      900
      960
```

3.
```
        7
      50r. 2
   5 )287
      250
       37
       35
        2
```

4.
```
        5
     ×57
      35
     250
     285 +2 = 287
```

5.
```
         7   ¹2
    5,⁸ ³ ¹2
   -2, 1 7 4
    3, 6 5 8
```

6.
```
      ⁸    ⁵
    ⁹,⁰ ⁶ ¹7
   -   1 5 8
    8, 9 0 9
```

7.
```
         ⁵
    4,9 ⁶ ¹5
   -3,4 5 6
    1,5 0 9
```

8. $20 \div 4 = 5$

9. $32 \div 4 = 8$

10. $16 \div 2 = 8$

11. true

12. 3,400,000,000,800

13. $639 \div 3 = 213$ yd

14. $3 \times 5,280 = 15,840$ ft

15. Kimberly: $560 + 237 = 797$
    $850 - 797 = 53$ words

16. $3 \times 2,000 = 6,000$
    $6,000 > 3,500$
    3 tons is better

Systematic Review19F

1.
```
       4
      00
    200r. 1
  4|817
    800
     17
     16
      1
```

2.
```
       4
    ×204
      16
      00
     800
   816 +1= 817
```

3.
```
       1
      00
     200
  2|402
     400
       2
       2
       0
```

4.
```
       2
    ×201
       2
      00
     400
     402
```

5. done

6.
```
      461
    ×  85
        3
     ¹2 005
        4
     3 288
     3 9,185
```

7.
```
      558
    ×  39
      ¹47
     ¹4 552
      1 2
     ¹1 554
     21,762
```

8. $100 \times 3 = 300$
9. $4 \times 5,280 = 21,120$
10. $17 \times 16 = 272$
11. perpendicular
12. $700 \div 7 = 100$
13. $1,251,621$
14. $131+155+100+102 = 488$
    $488 \div 4 = \$122$
15. $11 \times 10 = 110$
    $110 - 17 = 93$ jars of jelly

Lesson Practice 20A

1. done
2. done

3.
```
       50 4/7        7
     7|354         × 50
       350          350
         4          r.4
                    354
```

4.
```
       100 5/9        9
     9|905          ×100
       900          900
         5          r.5
                    905
```

5.
```
        70          6
     6|420         ×70
       420          420
         0
```

6.
```
       121 6/8        8
     8|974          ×121
       800            8
       174          160
       160          800
        14          968
         8          r.6
         6          974
```

7. $\$435 \div 2 = \$217\frac{1}{2}$
8. $185 \div 3 = 61\frac{2}{3}$ yd

Lesson Practice 20B

1.
$$41\tfrac{4}{6}$$
$$6\overline{)250}$$
240
10
6
4

6
×41
6
240
246
r.4
250

2.
$$34\tfrac{6}{8}$$
$$8\overline{)278}$$
240
38
32
6

8
×34
32
240
272
r.6
278

3.
$$53\tfrac{2}{4}$$
$$4\overline{)214}$$
200
14
12
2

4
×53
12
200
212
r.2
214

4.
$$116\tfrac{5}{7}$$
$$7\overline{)817}$$
700
117
70
47
42
5

7
×116
42
70
700
812
r.5
817

5.
$$109\tfrac{4}{6}$$
$$6\overline{)658}$$
600
58
54
4

6
×109
54
00
600
654
r.4
658

6.
$$12\tfrac{7}{9}$$
$$9\overline{)115}$$
90
25
18
7

9
×12
18
90
108
r.7
115

7. $145 ÷ $7 = $20\tfrac{5}{7}$ weeks

8. 253 ÷ 2 = $126\tfrac{1}{2}$ days

Lesson Practice 20C

1.
$$95\tfrac{2}{8}$$
$$8\overline{)762}$$
720
42
40
2

8
×95
40
720
760
r.2
762

2.
$$52\tfrac{3}{6}$$
$$6\overline{)315}$$
300
15
12
3

6
×52
12
300
312
r.3
315

3.
$$33\tfrac{5}{7}$$
$$7\overline{)236}$$
210
26
21
5

7
×33
21
210
231
r.5
236

4.
$$27\tfrac{3}{5}$$
$$5\overline{)138}$$
100
38
35
3

5
×27
35
100
135
r.3
138

5.
$$170\tfrac{1}{2}$$
$$2\overline{)341}$$
200
141
140
1

2
×170
0
140
200
340
r.1
341

6.
$$3\overline{)104} = 34\frac{2}{3}$$
$$\underline{90}$$
$$14$$
$$\underline{12}$$
$$2$$

$$\begin{array}{r} 3 \\ \times 34 \\ \hline 12 \\ 90 \\ \hline 102 \\ r.2 \\ \hline 104 \end{array}$$

7. $\$124 \div 8 = \$15\frac{4}{8}$

8. $275 \div 4 = 68\frac{3}{4}$ pails

9. $15 \div 6 = 2\frac{3}{6}$ pieces

10. $8 \times 5{,}280 = 42{,}240$ ft

11. perpendicular

12. $35 \times 2{,}000 = 70{,}000$ lb

13. $63 + 79 + 72 = 214$

$214 \div 3 = 71\frac{1}{3}$ degrees

14. $41 + 52 + 60 = 153$

$153 \div 3 = 51$ degrees

15. $1{,}912 - 1{,}787 = 125$ years

Systematic Review 20D

1.
$$6\overline{)847} = 141\frac{1}{6}$$
$$\underline{600}$$
$$247$$
$$\underline{240}$$
$$7$$
$$\underline{6}$$
$$1$$

2.
$$\begin{array}{r} 6 \\ \times 141 \\ \hline 6 \\ 240 \\ 600 \\ \hline 846 \\ r.1 \\ \hline 847 \end{array}$$

3.
$$5\overline{)104} = 20\frac{4}{5}$$
$$\underline{100}$$
$$4$$

4.
$$\begin{array}{r} 5 \\ \times 20 \\ \hline 100 \\ r.4 \\ \hline 104 \end{array}$$

5. done

6. $16 \div 4 = 4$ ft

7. $14 \div 7 = 2$ in

8. $925 \div 2 = 462\frac{1}{2}$ hours

Systematic Review 20E

1.
$$8\overline{)169} = 21\frac{1}{8}$$
$$\underline{160}$$
$$9$$
$$\underline{8}$$
$$1$$

2.
$$\begin{array}{r} 8 \\ \times 21 \\ \hline 8 \\ 160 \\ \hline 168 \\ r.1 \\ \hline 169 \end{array}$$

3.
$$7\overline{)724} = 103\frac{3}{7}$$
$$\underline{700}$$
$$24$$
$$\underline{21}$$
$$3$$

4.
$$\begin{array}{r} 7 \\ \times 103 \\ \hline 21 \\ 00 \\ 700 \\ \hline 721 \\ r.3 \\ \hline 724 \end{array}$$

5. $50 \div 5 = 10$ in

6. $15 \div 5 = 3$ ft

7. $72 \div 9 = 8$ in
8. $45 \div 9 = 5$ minutes
9. $67 \div 5 = 13$ r. 2; 13 packages, 2 stamps left over
10. $\$350 \div 8 = \$43\frac{6}{8}$
11. $2{,}000 \times 16 = 32{,}000$ ounces
12. $486 \div 4 = 121\frac{2}{4}$ dollars
13. yes
14. $1{,}959 - 1{,}788 = 171$ years
15. $60 \times 24 = 1{,}440$ minutes

Systematic Review 20F

1.
$$\begin{array}{r} 181\frac{1}{4} \\ 4\overline{)725} \\ 400 \\ \hline 325 \\ 320 \\ \hline 5 \\ 4 \\ \hline 1 \end{array}$$

2.
$$\begin{array}{r} 4 \\ \times 181 \\ \hline 4 \\ 320 \\ 400 \\ \hline 724 \\ r.1 \\ \hline 725 \end{array}$$

3.
$$\begin{array}{r} 51\frac{8}{9} \\ 9\overline{)467} \\ 450 \\ \hline 17 \\ 9 \\ \hline 8 \end{array}$$

4.
$$\begin{array}{r} 9 \\ \times 51 \\ \hline 9 \\ 450 \\ \hline 459 \\ r.8 \\ \hline 467 \end{array}$$

5. $100 \div 9 = 11\frac{1}{9}$ in
6. $25 \div 6 = 4\frac{1}{6}$ ft
7. $59 \div 8 = 7\frac{3}{8}$ in
8. $528 \div 3 = 176$ lb
9. $444 \div 8 = 55\frac{4}{8}$ miles
10. $2 \times 5{,}280 = 10{,}560$ ft
    $10{,}560 > 10{,}000$; so Bethany ran farther
11. $\$315 \div 3 = \$105$
    $\$112 > \$105$; so $315 for 3 tons is cheaper
12. $10 \times 10 = 100$ sq. mi
13. 4
14. $2 \times 70 = 140$
    $160 - 140 = 20$ lb
15. $20 + 15 = 35$ lb

Lesson Practice 21A

1. 50
2. 20
3. 50
4. 600
5. 200
6. 900
7. 2,000
8. 3,000
9. 6,000
10. done
11. done
12.
$$\begin{array}{r} 200 \\ 2\overline{)500} \end{array}$$

13.
$$2 \overline{\smash{\big)}\,489} \; \; 244\tfrac{1}{2}$$
$$\underline{400}$$
$$89$$
$$\underline{80}$$
$$9$$
$$\underline{8}$$
$$1$$

14.
$$3 \overline{\smash{\big)}\,400} \; \; 100$$

15.
$$3 \overline{\smash{\big)}\,356} \; \; 118\tfrac{2}{3}$$
$$56$$
$$\underline{30}$$
$$26$$
$$\underline{24}$$
$$2$$

16. $400 \div 3 \approx 100$

$395 \div 3 = 131\tfrac{2}{3}$ yd

Lesson Practice 21B

1. 30
2. 70
3. 90
4. 400
5. 600
6. 700
7. 2,000
8. 3,000
9. 3,000
10.
$$2 \overline{\smash{\big)}\,700} \; \; 300$$

11.
$$2 \overline{\smash{\big)}\,735} \; \; 367\tfrac{1}{2}$$
$$\underline{600}$$
$$135$$
$$\underline{120}$$
$$15$$
$$\underline{14}$$
$$1$$

12. $500 \div 6 \approx 80$
13. $487 \div 6 = 81\tfrac{1}{6}$
14. $900 \div 9 = 100$
15. $921 \div 9 = 102\tfrac{3}{9}$
16. $900 \div 8 \approx 100$

$938 \div 8 = 117\tfrac{2}{8}$ gallons

Lesson Practice 21C

1. 20
2. 40
3. 60
4. 500
5. 700
6. 200
7. 2,000
8. 2,000
9. 9,000
10.
$$3 \overline{\smash{\big)}\,400} \; \; 100$$

11.
$$3 \overline{\smash{\big)}\,436} \; \; 145\tfrac{1}{3}$$
$$\underline{300}$$
$$136$$
$$\underline{120}$$
$$16$$
$$\underline{15}$$
$$1$$

12. $800 \div 2 = 400$
13. $751 \div 2 = 375\tfrac{1}{2}$
14. $800 \div 5 \approx 100$
15. $845 \div 5 = 169$
16. $800 \div 4 = 200$

$820 \div 4 = 205$ jugs

Systematic Review 21D

1. 80
2. 200
3. 4,000
4. $900 \div 3 = 300$
5. $912 \div 3 = 304$
6. $378 \div 7 = 54$
7. $7 \times 54 = 378$
8. $559 \div 9 = 62\frac{1}{9}$
9. $9 \times 62 = 558$
   $558 + 1 = 559$
10. $7 \times 100,000 + 5 \times 10,000$
11. $3 \times 1,000,000 + 4 \times 100,000$
12. $8 \times 1,000,000,000$
13. $24 \div 3 = 8$ baskets
14. $11 \times 5,280 = 58,080$ ft
15. $3 \times 5 = 15$ cards
16. $\$375 \div 4 = \$93\frac{3}{4}$

Systematic Review 21E

1. 50
2. 200
3. 7,000
4. $800 \div 5 \approx 100$
5. $776 \div 5 = 155\frac{1}{5}$
6. $225 \div 4 = 56\frac{1}{4}$
7. $4 \times 56 = 224$
   $224 + 1 = 225$
8. $628 \div 6 = 104\frac{4}{6}$
9. $6 \times 104 = 624$
   $624 + 4 = 628$
10. $365 + 498 = 863$
11. $863 - 125 = 738$
12. $671 \times 32 = 21,472$
13. $21 \div 7 = 3$ per day
14. $10 \times 2,000 = 20,000$ lb
15. $70 \div 7 = 10$ yards

16. $7 \times 3 = 21$ feet wide
    $10 \times 3 = 30$ feet long

Systematic Review 21F

1. 20
2. 100
3. 8,000
4. $1,000 \div 7 \approx 100$
5. $980 \div 7 = 140$
6. $463 \div 3 = 154\frac{1}{3}$
7. $3 \times 154 = 462$
   $462 + 1 = 463$
8. $336 \div 8 = 42$
9. $8 \times 42 = 336$
10. $1,345 + 601 = 1,946$
11. $4,532 - 2,055 = 2,477$
12. $2,891 \times 12 = 34,692$
13. $22 + 26 = 48$
    $48 \div 2 = 24$
    $24 \times 19 = 456$ sq in
14. $58 + 67 + 91 + 88 = 304$
    $304 \div 4 = 76$ pods
15. $484 \div 2 = 242$ quart jars
16. $15 \times 18 = 270$
    $270 \div 2 = 135$ sq ft

Lesson Practice 22A

1. done
2. $$13\overline{)968} \quad 74\frac{6}{13}$$
   $$\underline{910}$$
   $$58$$
   $$\underline{52}$$
   $$6$$
   $74 \times 13 = 962$
   $962 + 6 = 968$

3. $$\begin{array}{r} 65\frac{5}{12} \\ 12\overline{)785} \\ \underline{720} \\ 65 \\ \underline{60} \\ 5 \end{array}$$
$65\times 12=780$
$780+5=785$

4. $$\begin{array}{r} 23 \\ 21\overline{)483} \\ \underline{420} \\ 63 \\ \underline{63} \\ 0 \end{array}$$
$23\times 21=483$

5. $$\begin{array}{r} 58 \\ 11\overline{)638} \\ \underline{550} \\ 88 \\ \underline{88} \\ 0 \end{array}$$
$58\times 11=638$

6. $$\begin{array}{r} 25\frac{2}{15} \\ 15\overline{)377} \\ \underline{300} \\ 77 \\ \underline{75} \\ 2 \end{array}$$
$25\times 15=375$
$375+2=377$

7. $450\div 18=25$ people
8. $390\div 26=15$ days

Lesson Practice 22B

1. $$\begin{array}{r} 13 \\ 13\overline{)169} \\ \underline{130} \\ 39 \\ \underline{39} \\ 0 \end{array}$$
$13\times 13=169$

2. $$\begin{array}{r} 12\frac{14}{41} \\ 41\overline{)506} \\ \underline{410} \\ 96 \\ \underline{82} \\ 14 \end{array}$$
$12\times 41=492$
$492+14=506$

3. $$\begin{array}{r} 11\frac{57}{61} \\ 61\overline{)728} \\ \underline{610} \\ 118 \\ \underline{61} \\ 57 \end{array}$$
$11\times 61=671$
$671+57=728$

4. $$\begin{array}{r} 41 \\ 15\overline{)615} \\ \underline{600} \\ 15 \\ \underline{15} \\ 0 \end{array}$$
$41\times 15=615$

5. $$\begin{array}{r} 24\frac{20}{26} \\ 26\overline{)644} \\ \underline{520} \\ 124 \\ \underline{104} \\ 20 \end{array}$$
$24\times 26=624$
$624+20=644$

5.
$$
\begin{array}{r}
24 \frac{20}{26} \\
26\overline{)644} \\
\underline{520} \\
124 \\
\underline{104} \\
20
\end{array}
$$

$24 \times 26 = 624$

$624 + 20 = 644$

6.
$$
\begin{array}{r}
11 \frac{6}{75} \\
75\overline{)831} \\
\underline{750} \\
81 \\
\underline{75} \\
6
\end{array}
$$

$11 \times 75 = 825$

$825 + 6 = 831$

7. $315 \div 15 = 21$ bags

8. $\$732 \div 12 = \$61$

Lesson Practice 22C

1.
$$
\begin{array}{r}
22 \\
12\overline{)264} \\
\underline{240} \\
24 \\
\underline{24} \\
0
\end{array}
$$

$22 \times 12 = 264$

2.
$$
\begin{array}{r}
27 \frac{8}{14} \\
14\overline{)386} \\
\underline{280} \\
106 \\
\underline{98} \\
8
\end{array}
$$

$27 \times 14 = 378$

$378 + 8 = 386$

3.
$$
\begin{array}{r}
43 \\
22\overline{)946} \\
\underline{880} \\
66 \\
\underline{66} \\
0
\end{array}
$$

$43 \times 22 = 946$

4.
$$
\begin{array}{r}
20 \frac{1}{10} \\
10\overline{)201} \\
\underline{200} \\
1
\end{array}
$$

$20 \times 10 = 200$

$200 + 1 = 201$

5.
$$
\begin{array}{r}
10 \frac{3}{25} \\
25\overline{)253} \\
\underline{250} \\
3
\end{array}
$$

$10 \times 25 = 250$

$250 + 3 = 253$

6.
$$
\begin{array}{r}
13 \\
38\overline{)494} \\
\underline{380} \\
114 \\
\underline{114} \\
0
\end{array}
$$

$13 \times 38 = 494$

7. $735 \div 21 = 35$ trees

8. $\$143 \div \$11 = 13$ books

Systematic Review 22D

1.
$$
\begin{array}{r}
23 \\
12\overline{)276} \\
\underline{240} \\
36 \\
\underline{36} \\
0
\end{array}
$$

2. $23 \times 12 = 276$

3.
$$29\frac{1}{24}$$
$$24\overline{)697}$$
$$\underline{480}$$
$$217$$
$$\underline{216}$$
$$1$$

4. $29 \times 24 = 696$
$696 + 1 = 697$

5.
$$26$$
$$3\overline{)78}$$
$$\underline{60}$$
$$18$$
$$\underline{18}$$
$$0$$

6. $26 \times 3 = 78$

7.
$$221\frac{3}{4}$$
$$4\overline{)887}$$
$$\underline{800}$$
$$87$$
$$\underline{80}$$
$$7$$
$$\underline{4}$$
$$3$$

8. $221 \times 4 = 884$
$884 + 3 = 887$

9. $10 + 32 = 42$
$42 \div 2 = 21$
$21 \times 12 = 252$ sq ft

10. $28 \times 8 = 224$
$224 \div 2 = 112$ sq in

11. $40 \times 40 = 1,600$ sq mi

12. done

13. done

14. $120 \div 12 = 10$

15. $169 \div 12 = 14\frac{1}{12}$ ft

16. $25 \times 5,280 = 132,000$ ft

Systematic Review 22E

1.
$$14\frac{13}{17}$$
$$17\overline{)251}$$
$$\underline{170}$$
$$81$$
$$\underline{68}$$
$$13$$

2. $14 \times 17 = 238$
$238 + 13 = 251$

3.
$$16$$
$$41\overline{)656}$$
$$\underline{410}$$
$$246$$
$$\underline{246}$$
$$0$$

4. $16 \times 41 = 656$

5.
$$5\frac{5}{8}$$
$$8\overline{)45}$$
$$\underline{40}$$
$$5$$

6. $5 \times 8 = 40$
$40 + 5 = 45$

7.
$$59\frac{2}{9}$$
$$9\overline{)533}$$
$$\underline{450}$$
$$83$$
$$\underline{81}$$
$$2$$

8. $59 \times 9 = 531$
$531 + 2 = 533$

9. $9 + 29 = 38$
$38 \div 2 = 19$
$19 \times 11 = 209$ sq ft

10. $5 \times 16 = 80$
$80 \div 2 = 40$ sq in

11. $21 \times 32 = 672$ sq yd

12. $24 \div 12 = 2$ ft

13. $10 \times 12 = 120$ in

14. $180 \div 12 = 15$ ft

15. $576 \div 18 = 32$ miles per gallon

16. $6 \times 12 = 72$ in

17. $2 \times 2,000 = 4,000$ lb
18. $4,000 \times 16 = 64,000$ oz
     $64,000 - 32 = 63,968$ oz

14. $216 \div 12 = 18$ ft
15. $144 \div 12 = 12$ dozen
16. $36 \div 4 = \$9$
17. $16 \div 2 = 8$ times
18. $60 \times 10 = 600$ mi

Systematic Review 22F

1.
$$22 \overline{)484}$$
quotient 22
$$\underline{440}$$
$$44$$
$$\underline{44}$$
$$0$$

2. $22 \times 22 = 484$

3.
$$14 \overline{)950} \quad 67 \tfrac{12}{14}$$
$$\underline{840}$$
$$110$$
$$\underline{98}$$
$$12$$

4. $67 \times 14 = 938$
    $938 + 12 = 950$

5.
$$6 \overline{)39} \quad 6 \tfrac{3}{6}$$
$$\underline{36}$$
$$3$$

6. $6 \times 6 = 36$
    $36 + 3 = 39$

7.
$$7 \overline{)354} \quad 50 \tfrac{4}{7}$$
$$\underline{350}$$
$$4$$

8. $50 \times 7 = 350$
    $350 + 4 = 354$

9. $30 + 50 = 80$
    $80 \div 2 = 40$
    $40 \times 40 = 1,600$ sq ft

10. $11 \times 34 = 374$
    $374 \div 2 = 187$ sq in

11. $25 \times 9 = 225$ sq ft

12. $72 \div 12 = 6$ ft

13. $16 \times 12 = 192$ in

Lesson Practice 23A

1. done
2. done
3.
$$(5) \overline{)(1000)} \quad (200)$$

4.
$$5 \overline{)1090} \quad 218$$
$$\underline{1000}$$
$$90$$
$$\underline{50}$$
$$40$$
$$\underline{40}$$
$$0$$
    $218 \times 5 = 1,090$

5.
$$(7) \overline{)(9000)} \quad (1000)$$

6.
$$7 \overline{)9479} \quad 1354 \tfrac{1}{7}$$
$$\underline{7000}$$
$$2479$$
$$\underline{2100}$$
$$379$$
$$\underline{350}$$
$$29$$
$$\underline{28}$$
$$1$$
    $1,354 \times 7 = 9,478$
    $9,478 + 1 = 9,479$

7. $2,368 \div 2 = 1,184$ groups
8. $5,385 \div 5 = 1,077$ customers

Lesson Practice 23B

1.
$$\frac{(600)}{(9)\overline{)(6000)}}$$

2.
$$673\tfrac{6}{9}$$
$$9\overline{)6063}$$
$$\underline{5400}$$
$$663$$
$$\underline{630}$$
$$33$$
$$\underline{27}$$
$$6$$

$673 \times 9 = 6{,}057$

$6{,}057 + 6 = 6{,}063$

3.
$$\frac{(1000)}{(3)\overline{)(3000)}}$$

4.
$$861\tfrac{1}{3}$$
$$3\overline{)2584}$$
$$\underline{2400}$$
$$184$$
$$\underline{180}$$
$$4$$
$$\underline{3}$$
$$1$$

$861 \times 3 = 2{,}583$

$2{,}583 + 1 = 2{,}584$

5.
$$\frac{(2000)}{(2)\overline{)(5000)}}$$

6.
$$2375\tfrac{1}{2}$$
$$2\overline{)4751}$$
$$\underline{4000}$$
$$751$$
$$\underline{600}$$
$$151$$
$$\underline{140}$$
$$11$$
$$\underline{10}$$
$$1$$

$2{,}375 \times 2 = 4{,}750$

$4{,}750 + 1 = 4751$

7. $1{,}125 \div 9 = 125$ mothers

8. $2{,}268 \div 5 = 453\tfrac{3}{5}$ miles

Lesson Practice 23C

1.
$$\frac{(1000)}{(4)\overline{)(6000)}}$$

2.
$$1505\tfrac{2}{4}$$
$$4\overline{)6022}$$
$$\underline{4000}$$
$$2022$$
$$\underline{2000}$$
$$22$$
$$\underline{20}$$
$$2$$

$1{,}505 \times 4 = 6{,}020$

$6{,}020 + 2 = 6{,}022$

3.
$$\frac{(800)}{(7)\overline{)(6000)}}$$

4.
$$876\tfrac{2}{7}$$
$$7\overline{)6134}$$
$$\underline{5600}$$
$$534$$
$$\underline{490}$$
$$44$$
$$\underline{42}$$
$$2$$

$876 \times 7 = 6{,}132$

$6{,}132 + 2 = 6{,}134$

5.
$$\frac{(300)}{(9)\overline{)(3000)}}$$

6.
$$279\tfrac{5}{9}$$
$$9\overline{)2516}$$
$$\underline{1800}$$
$$716$$
$$\underline{630}$$
$$86$$
$$\underline{81}$$
$$5$$

$279 \times 9 = 2{,}511$

$2{,}511 + 5 = 2{,}516$

7. $9{,}936 \div 6 = 1{,}656$ ladybugs

8. $5{,}280 \div 4 = 1{,}320$

$1{,}320 \div 7 = 188\tfrac{4}{7}$ feet per day

Systematic Review 23D

1.
$$384\tfrac{2}{4}$$
$$4\overline{)1538}$$
$$\underline{1200}$$
$$338$$
$$\underline{320}$$
$$18$$
$$\underline{16}$$
$$2$$

2. $384 \times 4 = 1,536$
$1,536 + 2 = 1,538$

3.
$$1453\tfrac{1}{3}$$
$$3\overline{)4360}$$
$$\underline{3000}$$
$$1360$$
$$\underline{1200}$$
$$160$$
$$\underline{150}$$
$$10$$
$$\underline{9}$$
$$1$$

4. $1,453 \times 3 = 4,359$
$4,359 + 1 = 4,360$

5.
$$18$$
$$51\overline{)918}$$
$$\underline{510}$$
$$408$$
$$\underline{408}$$

6. $18 \times 51 = 918$

7.
$$11\tfrac{4}{38}$$
$$38\overline{)422}$$
$$\underline{380}$$
$$42$$
$$\underline{38}$$
$$4$$

8. $11 \times 38 = 418$
$418 + 4 = 422$

9. done

10.
$$3\,7\,36$$
$$\underline{\times\ \ \ 31}$$
$$^{13}7\,36$$
$$^{1}2\ \ 1$$
$$\underline{9\,1\,9\,8}$$
$$1\,1\,5,8\,1\,6$$

11.
$$2\,9\,45$$
$$\underline{\times\ \ \ 63}$$
$$^{1}2\,1\,1$$
$$6\,7\,2\,5$$
$$1\,^{1}5\,2\,3$$
$$\underline{2\,4\,4\,0}$$
$$1\,8\,5,5\,3\,5$$

12. $600 \div 3 = 200$ yd
13. $48 \div 2 = 24$ qt
14. $16 \div 4 = 4$ gal
15. no: $485 \div 4 = 121\tfrac{1}{4}$ cows
16. no

Systematic Review 23E

1.
$$250\tfrac{3}{8}$$
$$8\overline{)2003}$$
$$\underline{1600}$$
$$403$$
$$\underline{400}$$
$$3$$

2. $250 \times 8 = 2,000$
$2,000 + 3 = 2,003$

3.
$$692\tfrac{1}{2}$$
$$2\overline{)1385}$$
$$\underline{1200}$$
$$185$$
$$\underline{180}$$
$$5$$
$$\underline{4}$$
$$1$$

4. $692 \times 2 = 1,384$
$1,384 + 1 = 1,385$

5.
$$\frac{39}{12 \overline{)468}}$$
$$\underline{360}$$
$$108$$
$$\underline{108}$$
$$0$$

6. $39 \times 12 = 468$

7.
$$51\frac{6}{10}$$
$$10\overline{)516}$$
$$\underline{500}$$
$$16$$
$$\underline{10}$$
$$6$$

8. $51 \times 10 = 510$
$510 + 6 = 516$

9.
$$1348$$
$$\times \quad 29$$
$$^1 2\,^1 3\,7$$
$$9762$$
$$1$$
$$\underline{^1 2\,6\,8\,6}$$
$$39,092$$

10.
$$7539$$
$$\times \quad 12$$
$$1\quad 1$$
$$^1 4\,^0 6\,8$$
$$\underline{7539}$$
$$90,468$$

11.
$$2000$$
$$\times \quad 56$$
$$12000$$
$$\underline{10000}$$
$$112,000$$

12. $444 \div 4 = \$111$

13. $720 \div 12 = 60$ ft

14. $80 \div 16 = 5$ lb

15. $2,067 \div 3 = 689$ bugs per hour

16. $4 \times 7 = 28$ qt
$28 \times 52 = 1,456$ qt per year

Systematic Review 23F

1.
$$1255\frac{1}{3}$$
$$3\overline{)3766}$$
$$\underline{3000}$$
$$766$$
$$\underline{600}$$
$$166$$
$$\underline{150}$$
$$16$$
$$\underline{15}$$
$$1$$

2. $1,255 \times 3 = 3,765$
$3,765 + 1 = 3,766$

3.
$$1498\frac{1}{6}$$
$$6\overline{)8989}$$
$$\underline{6000}$$
$$2989$$
$$\underline{2400}$$
$$589$$
$$\underline{540}$$
$$49$$
$$\underline{48}$$
$$1$$

4. $1,498 \times 6 = 8,988$
$8,988 + 1 = 8,989$

5.
$$5\frac{45}{71}$$
$$71\overline{)400}$$
$$\underline{355}$$
$$45$$

6. $5 \times 71 = 355$
$355 + 45 = 400$

7.
$$8$$
$$44\overline{)352}$$
$$\underline{352}$$
$$0$$

8. $8 \times 44 = 352$

9.
$$
\begin{array}{r}
3445 \\
\times\ 93 \\
\hline
{}^1 11 \\
9225 \\
{}^1{}_2{}^2 3\ 34 \\
7665 \\
\hline
3\ 2\ 0,385
\end{array}
$$

10.
$$
\begin{array}{r}
2317 \\
\times\ 64 \\
\hline
{}^1\ 2 \\
8248 \\
{}^1\ 4 \\
12862 \\
\hline
148,288
\end{array}
$$

11.
$$
\begin{array}{r}
8912 \\
\times\ 25 \\
\hline
4\ 1 \\
{}^1 4\ 0{}^1 5\ 50 \\
{}^1\ 1 \\
6824 \\
\hline
2\ 2\,2,800
\end{array}
$$

12. $5,280 \div 3 = 1,760$ yd
13. $4 \times 2,000 = 8,000$ lb
14. $1,000 \times 12 = 12,000$ in
15. $1,760 \times 2 = 3,520$ yd
16. perpendicular

Lesson Practice 24A

1. done
2. done

3.
$$
\overset{(200)}{(20)\overline{)(4000)}}
$$

4.
$$
\begin{array}{r}
200 \\
19\overline{)3800} \\
3800
\end{array}
\qquad
\begin{array}{r}
19 \\
\times 200 \\
\hline
3800
\end{array}
$$

5.
$$
\overset{(100)}{(60)\overline{)(7000)}}
$$

6.
$$
\begin{array}{r}
125\ \frac{45}{55} \\
55\overline{)6920} \\
5500 \\
\hline
1420 \\
1100 \\
\hline
320 \\
275 \\
\hline
45
\end{array}
\qquad
\begin{array}{r}
55 \\
\times 125 \\
\hline
275 \\
110 \\
55 \\
\hline
6875 \\
+\ 45 \\
\hline
6,920
\end{array}
$$

7. $\$6,929 \div 41 = 169$ hours
8. $\$1,875 \div \$15 = 125$ coats

Lesson Practice 24B

1. done
2. done

3.
$$
\overset{(40)}{(90)\overline{)(4000)}}
$$

4.
$$
\begin{array}{r}
42\ \frac{41}{91} \\
91\overline{)3863} \\
3640 \\
\hline
223 \\
182 \\
\hline
41
\end{array}
\qquad
\begin{array}{r}
91 \\
\times 42 \\
\hline
182 \\
364 \\
\hline
3822 \\
+\ 41 \\
\hline
3,863
\end{array}
$$

5.
$$
\overset{(200)}{(30)\overline{)(6000)}}
$$

6.
$$
\begin{array}{r}
164\ \frac{16}{34} \\
34\overline{)5592} \\
3400 \\
\hline
2192 \\
2040 \\
\hline
152 \\
136 \\
\hline
16
\end{array}
\qquad
\begin{array}{r}
34 \\
\times 164 \\
\hline
136 \\
204 \\
34 \\
\hline
5576 \\
+\ 16 \\
\hline
5,592
\end{array}
$$

7. $4,275 \div 45 = 95$ boxes
8. $1,008 \div 14 = 72$ groups

Lesson Practice 24C

1.
$$\frac{(600)}{(10)\,|\,(6000)}$$

2.
$$\begin{array}{r} 509\frac{5}{12} \\ 12\,\overline{)6113} \\ 6000 \\ \hline 113 \\ 108 \\ \hline 5 \end{array}$$

$$\begin{array}{r} 12 \\ \times 509 \\ \hline 108 \\ 600 \\ \hline 6108 \\ +\quad 5 \\ \hline 6,113 \end{array}$$

3.
$$\frac{(50)}{(80)\,|\,(4000)}$$

4.
$$\begin{array}{r} 55\frac{20}{75} \\ 75\,\overline{)4145} \\ 3750 \\ \hline 395 \\ 375 \\ \hline 20 \end{array}$$

$$\begin{array}{r} 75 \\ \times 55 \\ \hline 375 \\ 375 \\ \hline 4125 \\ +\quad 20 \\ \hline 4,145 \end{array}$$

5.
$$\frac{(100)}{(50)\,|\,(7000)}$$

6.
$$\begin{array}{r} 152\frac{11}{46} \\ 46\,\overline{)7003} \\ 4600 \\ \hline 2403 \\ 2300 \\ \hline 103 \\ 92 \\ \hline 11 \end{array}$$

$$\begin{array}{r} 46 \\ \times 152 \\ \hline 92 \\ 230 \\ 46 \\ \hline 6992 \\ +\quad 11 \\ \hline 7,003 \end{array}$$

7. $7,000 \div 35 = \$200$

8. $9,360 \div 52 = 180$ balls per week

Systematic Review 24D

1.
$$\begin{array}{r} 222\frac{3}{25} \\ 25\,\overline{)5553} \\ 5000 \\ \hline 553 \\ 500 \\ \hline 53 \\ 50 \\ \hline 3 \end{array}$$

2.
$$\begin{array}{r} 25 \\ \times 222 \\ \hline 1 \\ 140 \\ 140 \\ 40 \\ \hline 5550 \\ +\quad 3 \\ \hline 5553 \end{array}$$

3.
$$\begin{array}{r} 95 \\ 63\,\overline{)5985} \\ 5670 \\ \hline 315 \\ 315 \\ \hline 0 \end{array}$$

4.
$$\begin{array}{r} 63 \\ \times 95 \\ \hline 315 \\ 20 \\ 547 \\ \hline 5985 \end{array}$$

5.
$$\begin{array}{r} 363\frac{6}{8} \\ 8\,\overline{)2910} \\ 2400 \\ \hline 510 \\ 480 \\ \hline 30 \\ 24 \\ \hline 6 \end{array}$$

6.
```
        8
    ×3 6 3
       2 4
      4 8
     2 4
    2 9¹0 4
  +     6
    2 9 1 0
```

7.
$$9\frac{9}{11}$$
```
11 )108
    99
     9
```

8.
```
    1 1
  ×   9
    9 9
  +   9
   1 0 8
```

9. done

10.
```
        3 2 4
      × 3 1 5 2
        6 4 8
      1 ¹1 2
    1 5 0 0
      3 2 4
    1
    ¹9⁶6 2
  1,0 2 1,2 4 8
```

11.
```
          7 9 3
        × 1 2 8 4
        ¹3 1
      ²2 8 6 2
        7 2
      ²5 6 2 4
    ²1 1
      4 8 6
    7 9 3
  1,0 1 8,2 1 2
```

12. $12 \times 14 = 168$ sq ft

13. $16 + 14 + 7 + 20 + 8 = 65$
    $65 \div 5 = 13$ oz < 1 lb

14. $7 \times 12 = 84$ in

Systematic Review 24E

1.
$$87\frac{33}{34}$$
```
34 )2991
    2720
     271
     238
      33
```

2.
```
        3 4
      ×  8 7
        2 2
      3 1 8
    2 4 2
    2 9¹5 8
  +    3 3
   2,9 9 1
```

3.
$$548\frac{5}{12}$$
```
12 )6581
    6000
     581
     480
     101
      96
       5
```

4.
```
        1 2
      × 5 4 8
        9 6
      4 8
    6 0
    6 5¹7 6
  +     5
   6,5 8 1
```

5.
$$926\frac{3}{5}$$
```
5 )4633
   4500
    133
    100
     33
     30
      3
```

6.
```
        5
    ×926
      30
      10
      45
    4630
    +  3
    4633
```

7.
$$27 \overline{)495} \quad 18\frac{9}{27}$$
```
    270
    225
    216
      9
```

8.
```
      27
    ×18
    216
     27
    486
    + 9
    495
```

9.
```
       162
    ×4195
       31
      500
     ¹5 1
      948
    ²1 6 2
    2
    448
    679,590
```

10.
```
        281
     ×2703
         2
       643
     ¹1 5
      4¹6 7
     1
     46 2
     759,543
```

11.
```
       437
    ×1588
      ¹2 5
    ¹3 2 46
      2 5
    ²3 2 46
    1 3
    2 0 5 5
    4 3 7
    6 93,956
```

12. $10 \times 16 = 160$
    $160 \div 2 = 80$ sq in

13. $2,266 \div 11 = 206$ miles

14. E, F, H, M, N or Z
    (possibly I, depending on style)

Systematic Review 24F

1.
$$14 \overline{)1613} \quad 115\frac{3}{14}$$
```
    1400
     213
     140
      73
      70
       3
```

2.
```
       14
    ×115
      70
      14
      14
    1610
    +  3
    1,613
```

3.
$$91 \overline{)1223} \quad 13\frac{40}{91}$$
```
     910
     313
     273
      40
```

4.
$$\begin{array}{r} 91 \\ \times 13 \\ \hline 273 \\ ^191 \\ \hline 1183 \\ +\ 40 \\ \hline 1{,}223 \end{array}$$

5.
$$1890\,\frac{2}{4}$$
$$4\,\overline{)7562}$$
$$\begin{array}{r} 4000 \\ \hline 3562 \\ 3200 \\ \hline 362 \\ 360 \\ \hline 2 \end{array}$$

6.
$$\begin{array}{r} 4 \\ \times 1890 \\ \hline 36 \\ 32 \\ 4 \\ \hline 7560 \\ +\ 2 \\ \hline 7{,}562 \end{array}$$

7.
$$22\,\frac{8}{13}6$$
$$13\,\overline{)294}$$
$$\begin{array}{r} 260 \\ \hline 34 \\ 26 \\ \hline 8 \end{array}$$

8.
$$\begin{array}{r} 13 \\ \times 22 \\ \hline 26 \\ 26 \\ \hline 2^186 \\ +\ 8 \\ \hline 294 \end{array}$$

9.
$$\begin{array}{r} 561 \\ \times 1361 \\ \hline ^156\ 1 \\ ^13 \\ 3066 \\ 1 \\ ^11583 \\ 56\ 1 \\ \hline 763{,}521 \end{array}$$

10.
$$\begin{array}{r} 946 \\ \times 3722 \\ \hline 1 \\ ^2{}^1882 \\ 1 \\ ^31\ 882 \\ ^16\ 2\ 4 \\ 3\ 8\ 2 \\ ^12\ 1\ 1 \\ 7\ 2\ 8 \\ \hline 3{,}521{,}012 \end{array}$$

11.
$$\begin{array}{r} 837 \\ \times 4693 \\ \hline 2 \\ ^2{}^2{}^14\ 9\ 1 \\ 2\ 6 \\ ^27\ 2\ 7\ 3 \\ 1\ 4 \\ ^24\ 8\ 8\ 2 \\ 1\ 2 \\ 3\ 2\ 2\ 8 \\ \hline 3{,}928{,}041 \end{array}$$

12. $5 + 7 = 12$
$12 \div 2 = 6$
$6 \times 12 = 72$ sq in

13. $20 \div 8 = 2$ r. 4; 2 full vans and 1
part load = 3 vans, no

14. E, F, H, L, T; possibly I and J

Lesson Practice 25A

1. done
2. done

3.
$$\frac{(1,000)}{(60)\,(60,000)}$$

4.
$$1,054\ \frac{14}{56}$$

$$56\,\overline{)59038}$$
$$\underline{56000}$$
$$3038$$
$$\underline{2800}$$
$$238$$
$$\underline{224}$$
$$14$$

$$56$$
$$\times\,1054$$
$$224$$
^{1230}
$$50$$
$$\underline{56}$$
$$59024$$
$$\underline{+\quad14}$$
$$59,038$$

5.
$$\frac{(1,000)}{(70)\,(70,000)}$$

6.
$$985$$
$$73\,\overline{)71905}$$
$$\underline{65700}$$
$$6205$$
$$\underline{5840}$$
$$365$$
$$\underline{365}$$
$$0$$

$$73$$
$$\times\,985$$
^{131}
^{15255}
^{16264}
$$37$$
$$71,905$$

7. $40,000 \div 20 = 2,000$

Lesson Practice 25B

1. done
2. done

3.
$$\frac{(1,000)}{(500)\,(800,000)}$$

4.
$$1637\ \frac{125}{503}$$

$$503\,\overline{)823536}$$
$$\underline{503000}$$
$$320536$$
$$\underline{301800}$$
$$18736$$
$$\underline{15090}$$
$$3646$$
$$\underline{3521}$$
$$125$$

$$503$$
$$\times\,1637$$
$$2$$
^{13501}
^{11509}
$$3018$$
$$\underline{503}$$
$$823,411$$
$$\underline{+\quad125}$$
$$823,536$$

5. $963,600 \div 365 = 2,640$ feet per day

6. $26,052 feet per day

7. $26,052 \div 52 = $501 per week

Lesson Practice 25C

1.
$$\frac{(2,000)}{(40)\,(80,000)}$$

2.
$$2160\ \frac{25}{35}$$

$$35\,\overline{)75625}$$
$$\underline{70000}$$
$$5625$$
$$\underline{3500}$$
$$2125$$
$$\underline{2100}$$
$$25$$

$$35$$
$$\times\,2160$$
^{130}
$$8$$
$$35$$
$$10$$
$$\underline{6}$$
$$75600$$
$$\underline{+\quad25}$$
$$75,625$$

3.
$$\frac{(1,000)}{(900)\,(900,000)}$$

4.
$$1085\ \frac{533}{858}$$

$$858\,\overline{)931463}$$
$$\underline{858000}$$
$$73463$$
$$\underline{68640}$$
$$4823$$
$$\underline{4290}$$
$$533$$

$$858$$
$$\times\,1085$$
$$4^1240$$
^{26465}
$$404$$
^{1858}
$$930930$$
$$\underline{+\quad533}$$
$$931,463$$

5. $12,560 \div 16 = 785$ lb

6. $900,000 \div 300 = 3,000$ days

Systematic Review 25D

1.
$$1668\frac{29}{38}$$
$$38\overline{)63413}$$
$$\underline{38000}$$
$$25413$$
$$\underline{22800}$$
$$2613$$
$$\underline{2280}$$
$$333$$
$$\underline{304}$$
$$29$$

2.
$$38$$
$$\times 1{,}668$$
$$^1 2\,6$$
$$^2 1\,4\,4\,4$$
$$^2 1\,4\,8\,8$$
$$3\,8\,8$$
$$\underline{8}$$
$$6\,3\,3\,8\,4$$
$$\underline{+\quad 29}$$
$$6\,3{,}4\,1\,3$$

3.
$$2766\frac{152}{357}$$
$$357\overline{)987614}$$
$$\underline{714000}$$
$$273614$$
$$\underline{249900}$$
$$23714$$
$$\underline{21420}$$
$$2294$$
$$\underline{2142}$$
$$152$$

4.
$$357$$
$$\times 2766$$
$$^2 1\,3\,4$$
$$8\,0\,2$$
$$^2 1\,3\,4$$
$$8\,0\,2$$
$$2\,3\,4$$
$$1\,5\,9$$
$$1\,1$$
$$6\,0\,4$$
$$\overline{9\,8\,7\,4\,6\,2}$$
$$\underline{+\quad 1\,5\,2}$$
$$9\,8\,7{,}6\,1\,4$$

5.
$$1032\frac{3}{9}$$
$$9\overline{)9291}$$
$$\underline{9000}$$
$$291$$
$$\underline{270}$$
$$21$$
$$\underline{18}$$
$$3$$

6.
$$9$$
$$\times 1032$$
$$18$$
$$27$$
$$\underline{9}$$
$$9\,288$$
$$\underline{+\quad 3}$$
$$9{,}291$$

7.
$$569\frac{3}{12}$$
$$12\overline{)6831}$$
$$\underline{6000}$$
$$831$$
$$\underline{720}$$
$$111$$
$$\underline{108}$$
$$3$$

8.
$$12$$
$$\times 569$$
$$1$$
$$^1 9\,8$$
$$1\,6\,2$$
$$\underline{5\,0}$$
$$6\,8\,2\,8$$
$$\underline{+\quad 3}$$
$$6{,}8\,3\,1$$

9. $525 \div 15 = 35$ in

10. $28 \div 7 = 4$ ft

11. $500 \div 25 = 20$ in

12. $500 \times 12 = 6{,}000$ mi

13. $4{,}800 \div 12 = 400$ mph

14. $4{,}800 + 6{,}000 = 10{,}800$ mi

Systematic Review 25E

1.
$$1094 \frac{60}{63}$$
$$63 \overline{)68982}$$
$$\underline{63000}$$
$$5982$$
$$\underline{5670}$$
$$312$$
$$\underline{252}$$
$$60$$

2.
$$63$$
$$\times 1094$$
^{121}
$$5242$$
$$47$$
$$\underline{63}$$
$$68922$$
$$\underline{+\quad 60}$$
$$68982$$

3.
$$2157 \frac{39}{321}$$
$$321 \overline{)692436}$$
$$\underline{642000}$$
$$50436$$
$$\underline{32100}$$
$$18336$$
$$\underline{16050}$$
$$2286$$
$$\underline{2247}$$
$$39$$

4.
$$321$$
$$\times 2157$$
$$21$$
^{11147}
$$505$$
$$321$$
$$\underline{642}$$
$$692397$$
$$\underline{+\qquad 39}$$
$$692,436$$

5.
$$1660$$
$$2 \overline{)3320}$$
$$\underline{2000}$$
$$1320$$
$$\underline{1200}$$
$$120$$
$$\underline{120}$$
$$0$$

6.
$$2$$
$$\times 1660$$
$$12$$
$$12$$
$$\underline{2}$$
$$3320$$

7.
$$183 \frac{39}{52}$$
$$52 \overline{)9555}$$
$$\underline{5200}$$
$$4355$$
$$\underline{4160}$$
$$195$$
$$\underline{156}$$
$$39$$

8.
$$52$$
$$\times 183$$
^{1156}
$$416$$
$$0$$
$$\underline{52}$$
$$9516$$
$$\underline{+\quad 39}$$
$$9,555$$

9. $860 \div 20 = 43$ in

10. $88 \div 11 = 8$ ft

11. $255 \div 17 = 15$ ft

12. $360 \times 2 = 720$ eggs
$720 \div 12 = 60$ dozen

13. $143 \div 11 = 13$ ft

14. $\$1,843 - \$792 = \$1,051$

Systematic Review 25F

1.
$$1437\frac{21}{31}$$

$$31\overline{)44568}$$
```
  31000
  13568
  12400
   1168
    930
    238
    217
     21
```

2.
```
        31
    × 1437
       217
        93
     1¹24
      31
    44547
   +   21
   44,568
```

3.
```
        600
  420 )252000
      252000
           0
```

4.
```
       420
     × 600
     12000
    240000
    25,2000
```

5.
$$710\frac{3}{8}$$

$$8\overline{)5683}$$
```
  5600
    83
    80
     3
```

6.
```
        8
    ×710
        8
      56
    5680
   +   3
    5683
```

7.
$$88\frac{12}{18}$$

$$18\overline{)1596}$$
```
   1440
    156
    144
     12
```

8.
```
        18
     × 88
        6
    ¹ ¹684
      84
    1584
   +  12
    1596
```

9. $110 \div 55 = 2$ in

10. $10,560 \div 132 = 80$ ft

11. $868 \div 31 = 28$ in

12. $560 \div 16 = 35$ lb

13. $2,000 \times 16 = 32,000$ oz

14. $543 + 324 + 480 = 1,347$
     $1,347 \div 3 = 449$ mi per day

Lesson Practice 26A

1. done
2. $(5 \times 4) \times 2 = 40$ cu units
3. $(3 \times 3) \times 2 = 18$ cu units
4. $(3 \times 4) \times 2 = 24$ cu units
5. $(6 \times 2) \times 3 = 36$ cu in
6. $(2 \times 8) \times 1 = 16$ cu ft
7. $(3 \times 3) \times 7 = 63$ cu in
8. $(3 \times 3) \times 3 = 27$ cu yd
9. $(10 \times 12) \times 4 = 480$ cu in
10. $(2 \times 3) \times 5 = 30$ cu ft

Lesson Practice 26B

1. $(2 \times 2) \times 2 = 8$ cu units
2. $(5 \times 2) \times 2 = 20$ cu units
3. $(8 \times 3) \times 4 = 96$ cu in
4. $(4 \times 10) \times 2 = 80$ cu ft
5. $(5 \times 2) \times 3 = 30$ cu in
6. $(3 \times 7) \times 2 = 42$ cu ft
7. $(2 \times 2) \times 6 = 24$ cu in
8. $(4 \times 4) \times 4 = 64$ cu yd
9. $(4 \times 3) \times 3 = 36$ cu yd
10. $(5 \times 6) \times 3 = 90$ blocks

Lesson Practice 26C

1. $(3 \times 2) \times 1 = 6$ cu units
2. $(4 \times 3) \times 3 = 36$ cu units
3. $(6 \times 3) \times 5 = 90$ cu in
4. $(4 \times 6) \times 3 = 72$ cu ft
5. $(6 \times 2) \times 4 = 48$ cu in
6. $(3 \times 8) \times 4 = 96$ cu ft
7. $(5 \times 5) \times 8 = 200$ cu in
8. $(9 \times 9) \times 9 = 729$ cu yd
9. $(4 \times 7) = 28$ gallons
10. $(5 \times 2) \times 1 = 10$ cu ft
    $10 \times 7 = 70$ gal

Systematic Review 26D

1. $12 \times 12 \times 12 = 1,728$ cu ft
2. $4 \times 9 \times 5 = 180$ cu in
3.
$$
\begin{array}{r}
1306 \tfrac{2}{5} \\
5 \overline{)6532} \\
\underline{5000} \\
1532 \\
\underline{1500} \\
32 \\
\underline{30} \\
2
\end{array}
$$
4.
$$
\begin{array}{r}
5 \\
\times 1306 \\
\hline
30 \\
15 \\
\underline{5} \\
6530 \\
\underline{+\quad 2} \\
6532
\end{array}
$$
5.
$$
\begin{array}{r}
2460 \tfrac{4}{36} \\
36 \overline{)88564} \\
\underline{72000} \\
16564 \\
\underline{14400} \\
2164 \\
\underline{2164} \\
4
\end{array}
$$
6.
$$
\begin{array}{r}
36 \\
\times 2460 \\
\hline
{}^1 3 \\
1286 \\
124 \\
\underline{62} \\
88560 \\
\underline{+\qquad 4} \\
88564
\end{array}
$$
7. $10,560 \div 3 = 3,520$ yd
8. $100 \times 3 = 300$ ft
9. $300 \times 12 = 3,600$ in
10. $35 \div 7 = 5$ cu ft
11. $84 \times 7 = 588$ gal
12. $10 \times 4 \times 5 = 200$ cu in
    $10 \times 6 \times 3 = 180$ cu in
    $200 > 180$

13. $32 ÷ 4 = $8
14. 576 ÷ 24 = 24 days
15. 10 × 16 = 160 oz

8. 400 ÷ 16 = 25 lb
9. 60 × 16 = 960 oz
10. 7 × 6 × 2 = 84 cu ft
   84 × 7 = 588 gal
11. 324 ÷ 4 = 81 fireflies
12. 15 × 16 = 240 oz
   240 > 179; Joe gained more
13. $549,600 ÷ 12 = $45,800 per month
14. $65,600 − $45,800 = $19,800
15. 10 × 8 = 80; 80 ÷ 2 = 40 sq in

Systematic Review 26E

1. 10 × 10 × 10 = 1,000 cu ft
2. 7 × 11 × 3 = 231 cu in
3.
```
      16
  18|288
     180
     108
     108
       0
```
4.
```
    18
  × 16
     4
  ¹168
    8
   288
```
5.
```
       1202  245/248
  248|298341
      248000
       50341
       49600
         741
         496
         245
```
6.
```
     248
   × 1202
       1
   ¹1 486
  ¹486
  248
  298¹0¹96
  +   245
  298,341
```
7. 8 × 2,000 = 16,000 lb

Systematic Review 26F

1. 16 × 5 × 4 = 320 cu ft
2. 10 × 11 × 3 = 330 cu in
3.
```
      2710
   6|16260
     12000
      4260
      4200
        60
        60
         0
```
4.
```
        6
  × 2710
        6
      42
   12
  16,260
```
5.
```
       7180  11/13
   13|93351
      91000
       2351
       1300
       1051
       1040
         11
```

6.
$$
\begin{array}{r}
13 \\
\times 7180 \\
\hline
2 \\
{}^{1}184 \\
2\ \ 3 \\
71 \\
\hline
93340 \\
+\quad 11 \\
\hline
93{,}351
\end{array}
$$

7. $18 \div 2 = 9$ qt
8. $75 \times 4 = 300$ quarters
9. $64 \times 4 = 256$ qt
10. $25{,}000 \div 5 = 5{,}000$ eggs
11. $10 \times 10 \times 5 = 500$ cu ft
12. $500 \times 7 = 3{,}500$ gal
13. $3{,}500 \times 8 = 28{,}000$ lb
14. yes
15. $14 + 8 = 22$
    $22 \div 2 = 11$
    $11 \times 10 = 110$ sq in

Lesson Practice 27A

1. done
2. 6 blocks
   2 equal parts
   count 1 part
   $\frac{1}{2}$ of 6 is 3
3. 10 blocks
   5 equal parts
   count 4 parts
   $\frac{4}{5}$ of 10 is 8
4. 9 blocks
   3 equal parts
   count 2 parts
   $\frac{2}{3}$ of 9 is 6
5. 8 blocks
   4 equal parts
   count 3 parts
   $\frac{3}{4}$ of 8 is 6

Lesson Practice 27B

1. done
2. $\frac{1}{2}$ of 18 is 9
3. $\frac{5}{6}$ of 12 is 10
4. $20 \div 5 = 4$
   $4 \times 3 = 12$
5. $6 \div 3 = 2$
   $2 \times 2 = 4$
6. $8 \div 2 = 4$
   $4 \times 1 = 4$
7. $6 \div 3 = 2$
   $2 \times 1 = 2$
8. $8 \div 4 = 2$
   $2 \times 1 = 2$
9. $10 \div 5 = 2$
   $2 \times 4 = 8$
10. $6 \div 2 = 3$
    $3 \times 1 = 3$
11. $12 \div 3 = 4$
    $4 \times 2 = 8$
12. $8 \div 2 = 4$
    $4 \times 1 = 4$
13. $12 \div 4 = 3$
    $3 \times 1 = 3$ months
14. $10 \div 2 = 5$
    $5 \times 1 = 5$ fingers
15. $10 \div 5 = 2$
    $2 \times 2 = 4$ bulbs

Lesson Practice 27C

1. $\frac{1}{3}$ of 9 is 3
2. $\frac{3}{5}$ of 10 is 6
3. $\frac{2}{4}$ of 12 is 6
4. $16 \div 4 = 4$
   $4 \times 2 = 8$
5. $10 \div 5 = 2$
   $2 \times 1 = 2$
6. $18 \div 6 = 3$
   $3 \times 5 = 15$

7. $12 \div 4 = 3$
    $3 \times 3 = 9$
8. $10 \div 5 = 2$
    $2 \times 3 = 6$
9. $4 \div 2 = 2$
    $2 \times 1 = 2$
10. $12 \div 3 = 4$
    $4 \times 1 = 4$
11. $12 \div 2 = 6$
    $6 \times 1 = 6$
12. $16 \div 8 = 2$
    $2 \times 7 = 14$
13. $14 \div 7 = 2$
    $2 \times 2 = 4$ roses
14. $20 \div 5 = 4$
    $4 \times 1 = 4$ bandages
15. $16 \div 4 = 4$
    $4 \times 3 = 12$ years

8.
```
         2322
    21 | 48762
         42000
          6762
          6300
           462
           420
            42
            42
             0
```
9. $2,322 \times 21 = 48,762$
10. $5,280 \times 2 = 10,560$
11. $2,000 \times 4 = 8,000$
12. $75 \times 12 = 900$
13. $18 \div 3 = 6$
    $6 \times 1 = 6$ girls
14. $3 \times 3 \times 4 = 36$ cu ft
    $36 \times 7 = 252$ gal
15. $252 \times 8 = 2,016$ lb

Systematic Review 27D

1. $6 \div 2 = 3$
    $3 \times 1 = 3$
2. $16 \div 4 = 4$
    $4 \times 3 = 12$
3. $20 \div 5 = 4$
    $4 \times 2 = 8$
4. $18 \times 6 \times 5 = 540$ cu ft
5. $12 \times 14 \times 2 = 336$ cu in
6.
```
          6003 3/4
      4 | 24015
          24000
             15
             12
              3
```
7. $6,003 \times 4 = 24,012$
    $24,012 + 3 = 24,015$ Ç

Systematic Review 27E

1. $12 \div 6 = 2$
    $2 \times 1 = 2$
2. $21 \div 7 = 3$
    $3 \times 4 = 12$
3. $16 \div 8 = 2$
    $2 \times 1 = 2$
4. $20 \times 11 \times 7 = 1,540$ cu ft
5. $6 \times 9 \times 1 = 54$ cu in
6.
```
          3278
      9 | 29502
          27000
           2502
           1800
            702
            630
             72
             72
              0
```
7. $3,278 \times 9 = 29,502$

8.
$$2031 \tfrac{3}{44}$$

```
44 |89367
    88000
    ─────
     1367
     1320
     ─────
       47
       44
     ─────
        3
```

9. $2{,}031 \times 44 = 89{,}364$
$89{,}364 + 3 = 89{,}367$

10. $5 \times 3 = 15$

11. $7 \times 2 = 14$

12. $20 \times 4 = 80$

13. $9 \div 9 = 1$
$1 \times 3 = 3$ boys

14. $24 \div 6 = 4$
$4 \times 1 = 4$ hours per day

15. $4 \times 60 = 240$ minutes

Systematic Review 27F

1. $20 \div 4 = 5$
$5 \times 3 = 15$

2. $12 \div 3 = 4$
$4 \times 1 = 4$

3. $14 \div 2 = 7$
$7 \times 1 = 7$

4. $31 \times 15 \times 8 = 3{,}720$ cu ft

5. $10 \times 12 \times 3 = 360$ cu in

6.
```
     3006
6 |18036
   18000
   ─────
      36
      36
   ─────
       0
```

7. $3{,}006 \times 6 = 18{,}036$

8.
```
      1417
25 |35425
    25000
    ─────
    10425
    10000
    ─────
      425
      250
    ─────
      175
      175
    ─────
        0
```

9. $1{,}417 \times 25 = 35{,}425$

10. $20 \times 16 = 320$

11. $35 \times 4 = 140$

12. $5{,}280 \times 3 = 15{,}840$

13. $8{,}227{,}332 + 3{,}305{,}006 +$
$4{,}884{,}234 = 16{,}416{,}572$ people

14. $8{,}274{,}961 - 7{,}477{,}503 = 797{,}458$ people

15. $8{,}274{,}961 + 7{,}477{,}503 = 15{,}752{,}464$
$15{,}752{,}464 - 14{,}445{,}000 =$
$1{,}307{,}464$ people

Lesson Practice 28A

1. done
2. 23
3. 17
4. 50
5. 200
6. 90
7. 70
8. 46
9. done
10. XIV
11. XXXI
12. XLVIII
13. CL
14. CCCXXV
15. CCXLIX
16. LXIII
17. 10:00
18. XIV
19. I, X, C
20. L, V

Lesson Practice 28B

1. 9
2. 36
3. 124
4. 152
5. 98
6. 192
7. 145
8. 315
9. XVIII
10. XXVI
11. XCIV
12. XLIII
13. CCLVIII
14. CCCXVII
15. CCLXII
16. CLXXXIX
17. 4:00
18. subtract
19. no
20. VI

Lesson Practice 28C

1. 19
2. 49
3. 94
4. 53
5. 147
6. 18
7. 39
8. 222
9. XCIV
10. XIII
11. XXVIII
12. XLV
13. CCCLIX
14. CVIII
15. CCCXI
16. CCLXXIV
17. 9:00
18. 20

19. 6:00
20. 3

Systematic Review 28D

1. 8
2. 24
3. 140
4. 42
5. LXI
6. XLVIII
7. CLII
8. CCX
9. $12 \div 3 = 4$
   $4 \times 1 = 4$
10. $21 \div 7 = 3$
    $3 \times 2 = 6$
11. $10 \div 5 = 2$
    $2 \times 4 = 8$
12. $12 \times 45 = 540$
    $540 \div 2 = 270$ sq in
13.
$$
\begin{array}{r}
8763\frac{4}{7} \\
7\overline{)61345} \\
\underline{56000} \\
5345 \\
\underline{4900} \\
445 \\
\underline{420} \\
25 \\
\underline{21} \\
4
\end{array}
$$
14. $8,763 \times 7 = 61,341$
    $61,341 + 4 = 61,345$
15.
$$
\begin{array}{r}
2014 \\
425\overline{)855950} \\
\underline{850000} \\
5950 \\
\underline{4250} \\
1700 \\
\underline{1700} \\
0
\end{array}
$$
16. $2,014 \times 425 = 855,950$

17. $24 \div 3 = 8$
$8 \times 1 = 8$ children

18. $7 + 9 + 11 + 1 = 28$
$28 \div 4 = 7$ books

Systematic Review 28E

1. 24
2. 271
3. 145
4. 18
5. LXXV
6. XCII
7. CCCLXXX
8. CXI
9. $16 \div 4 = 4$
$4 \times 1 = 4$
10. $18 \div 6 = 3$
$3 \times 5 = 15$
11. $20 \div 2 = 10$
$10 \times 1 = 10$
12. $8 + 14 = 22$
$22 \div 2 = 11$
$11 \times 11 = 121$ sq ft
13.
$$\begin{array}{r} 1528\frac{1}{9} \\ 9\overline{)13753} \\ \underline{9000} \\ 4753 \\ \underline{4500} \\ 253 \\ \underline{180} \\ 73 \\ \underline{72} \\ 1 \end{array}$$
14. $1{,}528 \times 9 = 13{,}752$
$13{,}752 + 1 = 13{,}753$

15.
$$\begin{array}{r} 2241\frac{246}{350} \\ 350\overline{)784596} \\ \underline{700000} \\ 84596 \\ \underline{70000} \\ 14596 \\ \underline{14000} \\ 596 \\ \underline{350} \\ 246 \end{array}$$
16. $350 \times 2241 = 784{,}350$
$784{,}350 + 246 = 784{,}596$
17. $12 + 11 + 2 + 3 = 28$
$28 \div 4 = 7$ in
18. $5 \times 12 = 60$
$60 + 7 = 67$ in

Systematic Review 28F

1. 140
2. 354
3. 27
4. 81
5. XXXIV
6. LVI
7. CCXCIX
8. CCCLV
9. $9 \div 3 = 3$
$3 \times 2 = 6$
10. $15 \div 5 = 3$
$3 \times 3 = 9$
11. $12 \div 4 = 3$
$3 \times 1 = 3$
12. $65 \times 65 = 4{,}225$ sq mi

13.
```
        11197
    3 | 33591
        30000
         3591
         3000
          591
          300
          291
          270
           21
           21
            0
```

14. $11,197 \times 3 = 33,591$

15.
```
         13330  14
                ──
                48
    48 | 639854
         480000
         159854
         144000
          15854
          14400
           1454
           1440
             14
```

16. $13,330 \times 48 = 639,840$
    $639,840 + 14 = 639,854$

17. parallel

18. $12 \div 4 = 3$
    $3 \times 3 = 9$ months

Lesson Practice 29A

1. done

2. $\dfrac{1}{3}$

3. $\dfrac{4}{5}$

4. $\dfrac{1}{2}$

5. $\dfrac{3}{4}$

6. $\dfrac{1}{6}$

7. $\dfrac{2}{5}$

8. $\dfrac{2}{3}$

9. denominator is 6, so 6 pieces

Lesson Practice 29B

1. done

2.

3.

4.

5.

6.

7.

8.

9. numerator is 1, so 1 piece

7.

8.

9. numerator is 3, so 3 pieces

Lesson Practice 29C

1. done

2.

3.

4.

5.

6.

Systematic Review 29D

1. $\frac{3}{6}$

2. $\frac{1}{4}$

3. 9

4. 17

5. 250

6. 99

7. XCI

8. LIV

9. LXIII

10. CCCXCI

11. $12 \times 32 = 384$ sq in

12. $21 \times 40 = 840$
    $840 \div 2 = 420$ sq in

13.
$$116 \frac{6}{17}$$
$$17 \overline{)1978}$$
$$\underline{1700}$$
$$278$$
$$\underline{170}$$
$$108$$
$$\underline{102}$$
$$6$$

14. $116 \times 17 = 1,972$
    $1,972 + 6 = 1,978$

15.
```
        1061
   38 ⌐40318
        38000
         2318
         2280
           38
           38
            0
```

16. $1{,}061 \times 38 = 40{,}318$

17. perpendicular

18. $15 \div 3 = 5$
    $1 \times 5 = 5$ jelly beans

Systematic Review 29E

1.

2.

3. 72
4. 34
5. 194
6. 65
7. XXV
8. XIII
9. XLVIII
10. CLX
11. $16 \times 10 \times 5 = 800$ cu ft
12. $72 \times 91 \times 23 = 150{,}696$ cu in
13.
```
        53 14/65
   65 ⌐3459
        3250
         209
         195
          14
```

14. $65 \times 53 = 3{,}445$
    $3{,}445 + 14 = 3{,}459$

15.
```
        2648 19/21
   21 ⌐55627
        42000
        13627
        12600
         1027
          840
          187
          168
           19
```

16. $2{,}648 \times 21 = 55{,}608$
    $55{,}608 + 19 = 55{,}627$

17. $200 \div 50 = 4$ hours

18. $12 \div 4 = 3$
    $3 \times 3 = 9$ for Naomi
    $12 - 9 = 3$ for Ruth

Systematic Review 29F

1.

2.

3. 14
4. 86
5. 209
6. 198
7. XXXVI
8. XIV
9. LIX
10. CCLXXIII
11. $100 \times 50 \times 30 = 150{,}000$ cu ft
12. $19 \times 24 \times 9 = 4{,}104$ cu in

13.
$$182 \frac{38}{42}$$

```
    182 38/42
42 )7682
    4200
    3482
    3360
     122
      84
      38
```

14. $182 \times 42 = 7,644$
$7,644 + 38 = 7,682$

15.
```
      2400
25 )60000
    50000
    10000
    10000
        0
```

16. $2,400 \times 25 = 60,000$

17. $350 \div 60 = 5 \frac{50}{60}$; 5 hours, 50 minutes

18. $50 \div 2 = 25$
$25 \times 1 = 25$ people

Lesson Practice 30A

1. done
2. done
3. 400
4. 1,250
5. 45,000
6. 952
7. 700
8. 2,003
9. done
10. done
11. DLXXVIII
12. $\overline{\text{V}}$
13. MMCXLVI
14. MDCCCLXXII
15. $\overline{\text{X}}$
16. MMLXV
17. 1945
18. MCMLV

19. MDCCLXXVI
20. $\overline{\text{V}}$

Lesson Practice 30B

1. 556
2. 1,421
3. 549
4. 5,200
5. 931
6. 514
7. 2,000,000
8. 1,320
9. DI
10. DCXXV
11. $\overline{\text{C}}$
12. MMM
13. $\overline{\text{CM}}$
14. CDXXXII
15. MCCLXIII
16. MCMLXVII
17. 1861
18. DLXIII
19. MCDXCII
20. 1,555

Lesson Practice 30C

1. 584
2. 606
3. 945
4. 20,000
5. 1,140
6. 829
7. 50,000
8. 900,000
9. DLXXXII
10. CMLXXIII
11. MLIII
12. MMMCC
13. CDXLIV
14. MDX
15. MMMCMXCV

16. $\overline{\text{V}}$
17. 1799
18. MCMLXVI
19. CDLIX
20. 1066

Systematic Review 30D

1. 2,900
2. 534
3. 1,600
4. 955
5. LVII
6. CIX
7. DLXXX
8. MCDXI
9. $\dfrac{4}{5}$
10. $\dfrac{2}{3}$
11.
$$
\begin{array}{r}
101\frac{23}{89} \\
89\overline{)9012} \\
8900 \\
\hline
112 \\
89 \\
\hline
23
\end{array}
$$
12. $101 \times 89 = 8,989$
$8,989 + 23 = 9,012$
13.
$$
\begin{array}{r}
2010 \\
16\overline{)32160} \\
32000 \\
\hline
160 \\
160 \\
\hline
0
\end{array}
$$
14. $2,010 \times 16 = 32,160$
15. perpendicular
16. $8 \times 9 = 72$
$72 \div 2 = 36$ sq ft
17. $85 + 92 + 73 + 98 = 348$
$348 \div 4 = 87$
18. $\$345 = \$128 = \$217$

Systematic Review 30E

1. 928
2. 414
3. 1,000,000
4. 80
5. XXIX
6. CCXCIX
7. MMCMXCIX
8. $\overline{\text{V}}$
9. $\dfrac{2}{5}$
10. $\dfrac{1}{2}$
11.
$$
\begin{array}{r}
208\frac{1}{38} \\
38\overline{)7905} \\
7600 \\
\hline
305 \\
304 \\
\hline
1
\end{array}
$$
12. $208 \times 38 = 7,904$
$7,904 + 1 = 7,905$
13.
$$
\begin{array}{r}
2296\frac{126}{200} \\
200\overline{)459326} \\
400000 \\
\hline
59326 \\
40000 \\
\hline
19326 \\
18000 \\
\hline
1326 \\
1200 \\
\hline
126
\end{array}
$$
14. $2,296 \times 200 = 459,200$
$459,200 + 126 = 459,326$
15. $17 \times 4 = 68$ qt
16. $5 \times 4 = 20$ quarters
17. $15 \times 21 = 315$ sq ft
18. $32 \div 16 = 2$ lb

Systematic Review 30F

1. 2,310
2. 590
3. 50,000
4. 171
5. XVII
6. CDL
7. MMMCCLXIX
8. DCLXXII
9. $1333\frac{49}{51}$

$$51\overline{)68032}$$
$$\underline{51000}$$
$$17032$$
$$\underline{15300}$$
$$1732$$
$$\underline{1530}$$
$$202$$
$$\underline{153}$$
$$49$$

$1{,}333 \times 51 = 67{,}983$
$67{,}983 + 49 = 68{,}032$

10. $3143\frac{83}{182}$

$$182\overline{)572109}$$
$$\underline{546000}$$
$$26109$$
$$\underline{18200}$$
$$7909$$
$$\underline{7280}$$
$$629$$
$$\underline{546}$$
$$83$$

$3{,}143 \times 182 = 572{,}026$
$572{,}026 + 83 = 572{,}109$

11. $4 \times 2{,}000 = 8{,}000$ lb
12. $6 + 8 = 14$
$14 \div 2 = 7$
$7 \times 5 = 35$ sq ft
13. $24 \div 3 = 8$
$8 \times 2 = 16$ swallows
14. $5{,}280 \times 2 = 10{,}560$ ft
15. 3,000
16. $5 \times 3 = 15$ ft.
$15 \times 12 = 180$ in
17. $315 \div 45 = 7$ hours
18. $5 \times 6 \times 3 = 90$ cu ft

Test Solutions

Test 1

1. $4 \times 4 = 16$
2. $5 \times 3 = 15$
   $3 \times 5 = 15$
3. $3 \times \underline{9} = 27$
4. $8 \times \underline{8} = 64$
5. $4 \times \underline{5} = 20$
6. $6 \times \underline{1} = 6$
7. $5 \times \underline{9} = 45$
8. $6 \times \underline{4} = 24$
9. $10 \times \underline{10} = 100$
10. $2 \times \underline{7} = 14$
11. $5 \times 5 = 25$ sq ft
12. $8 \times 10 = 80$ sq in
13. $9 \times 7 = 63$ sq mi
14. $9 \times \underline{10} = \$90$; 10 hours
15. $8 \times 9 = 72$ sq ft

Test 2

1. $8 \div 1 = 8$
2. $18 \div 2 = 9$
3. $6 \div 2 = 3$
4. $4 \div 1 = 4$
5. $10 \div 2 = 5$
6. $8 \div 2 = 4$
7. $\dfrac{2}{2} = 1$
8. $\dfrac{7}{1} = 7$
9. $\dfrac{4}{2} = 2$
10. $5 \times \underline{3} = 15$
11. $5 \times \underline{5} = 25$
12. $2 \times \underline{10} = 20$
13. $10 \times \underline{6} = 60$
14. $8 \times 5 = 40$
15. $10 \times 7 = 70$
16. $4 \times 5 = 20$

Test 3

17. $10 \times 3 = 30$
18. $10 \times 8 = 80$ sq in
19. $12 \div 2 = 6$ gumdrops
20. $8 \div 8 = 1$ piece each

1. $20 \div 10 = 2$
2. $80 \div 10 = 8$
3. $60 \div 10 = 6$
4. $90 \div 10 = 9$
5. $8 \div 1 = 8$
6. $14 \div 2 = 7$
7. $16 \div 2 = 8$
8. $5 \div 1 = 5$
9. $6 \div 2 = 3$
10. $10 \div 2 = 5$
11. $10 \div 10 = 1$
12. $4 \div 2 = 2$
13. $5 \times 3 = 15$
14. $3 \times 3 = 9$
15. $6 \times 3 = 18$
16. $3 \times 4 = 12$
17. $10 \times 2 = 20$ pints
18. $8 \times 3 = 24$ ft
19. $\$40 \div \$10 = 4$ gifts
20. $7 \times 9 = 63$ sq in

Test 4

1. $15 \div 5 = 3$
2. $18 \div 3 = 6$
3. $12 \div 3 = 4$
4. $45 \div 5 = 9$
5. $30 \div 5 = 6$
6. $25 \div 5 = 5$
7. $27 \div 3 = 9$
8. $6 \div 3 = 2$
9. $14 \div 2 = 7$

10. $50 \div 10 = 5$
11. $24 \div 3 = 8$
12. $9 \div 3 = 3$
13. $3 \times \underline{6} = 18$
14. $7 \times \underline{3} = 21$
15. $5 \times \underline{4} = 20$

16.
$$\begin{array}{r} {}^1 2\,5 \\ +\,3\,8 \\ \hline 6\,3 \end{array}$$

17.
$$\begin{array}{r} {}^1 4\,7 \\ +\,7\,3 \\ \hline 1\,2\,0 \end{array}$$

18.
$$\begin{array}{r} {}^1 6\,4 \\ +\,1\,9 \\ \hline 8\,3 \end{array}$$

19. $5 \times 7 = 35$ sq mi
20. $75 + 69 = 144$ beads

Test 5

1. $9 \times \underline{9} = 81$
2. $9 \times \underline{2} = 18$
3. $9 \times \underline{5} = 45$
4. $9 \times \underline{8} = 72$
5. $9 \times \underline{6} = 54$
6. $9 \times \underline{3} = 27$
7. $9 \times \underline{7} = 63$
8. $9 \times \underline{4} = 36$
9. $35 \div 5 = 7$
10. $24 \div 3 = 8$
11. $5 \div 5 = 1$
12. $18 \div 3 = 6$
13. $21 \div 3 = 7$
14. $25 \div 5 = 5$
15. $15 \div 3 = 5$
16. $16 \div 2 = 8$
17. $\parallel$
18. $\perp$
19. $30 \div 3 = 10$ yd
20. $\$15 + \$26 = \$41$

Test 6

1. $18 \div 9 = 2$
2. $36 \div 9 = 4$
3. $45 \div 9 = 5$
4. $81 \div 9 = 9$
5. $27 \div 9 = 3$
6. $54 \div 9 = 6$
7. $72 \div 9 = 8$
8. $63 \div 9 = 7$
9. $90 \div 10 = 9$
10. $10 \div 2 = 5$
11. $15 \div 3 = 5$
12. $12 \div 3 = 4$
13. $24 \div 3 = 8$
14. $20 \div 2 = 10$
15. $9 \div 9 = 1$
16. $21 \div 3 = 7$

17.
$$\begin{array}{r} {}^8 9\,{}^1 1 \\ -\,7\,6 \\ \hline 1\,5 \end{array}$$

18.
$$\begin{array}{r} {}^3 4\,{}^1 2 \\ -\,1\,3 \\ \hline 2\,9 \end{array}$$

19.
$$\begin{array}{r} {}^7 8\,{}^1 0 \\ -\,3\,5 \\ \hline 4\,5 \end{array}$$

20. $18 + 17 = 35$
$35 \div 5 = 7$ treats each

Unit Test I

1. $4 \div 2 = 2$
2. $24 \div 3 = 8$
3. $45 \div 9 = 5$
4. $2 \div 1 = 2$
5. $18 \div 3 = 6$
6. $30 \div 5 = 6$
7. $81 \div 9 = 9$
8. $60 \div 10 = 6$
9. $16 \div 2 = 8$
10. $21 \div 3 = 7$

11. $63 \div 9 = 7$
12. $5 \div 5 = 1$
13. $27 \div 9 = 3$
14. $10 \div 2 = 5$
15. $40 \div 10 = 4$
16. $12 \div 3 = 4$
17. $45 \div 9 = 5$
18. $14 \div 2 = 7$
19. $25 \div 5 = 5$
20. $18 \div 2 = 9$
21. $50 \div 5 = 10$
22. $9 \div 3 = 3$
23. $9 \div 9 = 1$
24. $27 \div 3 = 9$
25. $90 \div 10 = 9$
26. $15 \div 3 = 5$
27. $12 \div 2 = 6$
28. $4 \div 1 = 4$
29. $40 \div 5 = 8$
30. $54 \div 9 = 6$
31. $70 \div 10 = 7$
32. $15 \div 5 = 3$
33. $45 \div 5 = 9$
34. $30 \div 3 = 10$
35. $5 \div 1 = 5$
36. $100 \div 10 = 10$
37. $9 \div 1 = 9$
38. $20 \div 5 = 4$
39. $72 \div 9 = 8$
40. $20 \div 2 = 10$
41. $6 \div 3 = 2$
42. $10 \div 5 = 2$
43. $18 \div 9 = 2$
44. $35 \div 5 = 7$
45. $8 \div 2 = 4$
46. $80 \div 10 = 8$
47. $8 \div 1 = 8$
48. $3 \div 3 = 1$
49. $56 + 39 = 95$

50. $62 - 25 = 37$
51. $81 - 46 = 35$
52. $\parallel$
53. $\perp$
54. $8 \times 9 = 72$ sq ft
55. $5 \times 3 = 15$ ft
56. $27 \div 3 = 9$ yd
57. $6 \times 2 = 12$ pt
58. $24 \div 2 = 12$ qt

Test 7

1. $9 \times 10 = 90$ sq ft
2. $3 \times 7 = 21$ sq in
3. $36 \div 9 = 4$
4. $18 \div 3 = 6$
5. $20 \div 5 = 4$
6. $12 \div 2 = 6$
7. $81 \div 9 = 9$
8. $40 \div 10 = 4$
9. $4 \div 2 = 2$
10. $63 \div 9 = 7$
11. $6 \times 2 = 12$
12. $6 \times 5 = 30$
13. $6 \times 9 = 54$
14. $6 \times 3 = 18$
15. $6 \times 7 = 42$
16. $6 \times 8 = 48$
17. $6 \times 4 = 24$
18. $6 \times 6 = 36$
19. $25 - 16 = 9$
    $9 \div 3 = 3$ cages
20. $18 \div 2 = 9$ qt

Test 8

1. $12 \div 6 = 2$
2. $24 \div 6 = 4$
3. $54 \div 6 = 9$
4. $30 \div 6 = 5$
5. $42 \div 6 = 7$
6. $48 \div 6 = 8$
7. $18 \div 6 = 3$
8. $36 \div 6 = 6$
9. $72 \div 9 = 8$
10. $20 \div 5 = 4$
11. $8 \div 2 = 4$
12. $27 \div 3 = 9$
13.
$$\begin{array}{r} {}^1\cancel{2}\,{}^1 3 \\ -\ \ 5 \\ \hline 1\ 8 \end{array}$$
14.
$$\begin{array}{r} {}^1 7\ 2 \\ +\ 1\ 9 \\ \hline 9\ 1 \end{array}$$
15.
$$\begin{array}{r} {}^4\cancel{5}\,{}^1 3 \\ -\ 4\ 5 \\ \hline 8 \end{array}$$
16.
$$\begin{array}{r} 2\ 2 \\ \times 1\ 3 \\ \hline 6\ 6 \\ 2\ 2\ \ \\ \hline 2\ 8\ 6 \end{array}$$
17.
$$\begin{array}{r} 4\ 5 \\ \times 2\ 4 \\ \hline {}^1 2 \\ 1\ 6\ 0 \\ 8\ 0\ \ \\ \hline 1,0\ 8\ 0 \end{array}$$
18.
$$\begin{array}{r} 1\ 6 \\ \times 3\ 7 \\ \hline {}^1 4 \\ 1\ 7\ 2 \\ 3\ 8\ \ \\ \hline 5\ 9\ 2 \end{array}$$
19. $20 \div 2 = 10$ people
20. $36 \div 6 = 6$ ft

Test 9

1. $4 \times 5 = 20$
   $20 \div 2 = 10$ sq in
2. $7 \times 6 = 42$ sq ft
3. $3 \times 2 = 6$
   $6 \div 2 = 3$ sq yd
4. $40 \div 10 = 4$
5. $12 \div 3 = 4$
6. $8 \div 2 = 4$
7. $45 \div 5 = 9$
8. $4 \times \underline{6} = 24$
9. $4 \times \underline{8} = 32$
10. $4 \times \underline{4} = 16$
11. $4 \times \underline{7} = 28$
12.
$$\begin{array}{r} 8\ 4 \\ \times 2\ 2 \\ \hline {}^1 1\ 6\ 8 \\ 1\ 6\ 8\ \ \\ \hline 1,8\ 4\ 8 \end{array}$$
13.
$$\begin{array}{r} 4\ 3 \\ \times 3\ 5 \\ \hline {}^1 2\ 1\ 5 \\ 1\ 2\ 9\ \ \\ \hline 1,5\ 0\ 5 \end{array}$$
14.
$$\begin{array}{r} 6\ 7 \\ \times 5\ 4 \\ \hline {}^2 \\ 2\ 4\ 8 \\ 3\ 3\ 5\ \ \\ \hline 3,6\ 1\ 8 \end{array}$$
15. $25 + 15 + 24 + 61 = 125$
16. $44 + 38 + 62 + 56 + 11 = 211$
17. $90 + 23 + 57 + 18 + 82 = 270$
18. no
19. $25 + 16 + 18 + 32 = 91$ animals
20. $91 \times 4 = 364$ hooves

Test 10

1. $8 \div 4 = 2$
2. $32 \div 4 = 8$
3. $16 \div 4 = 4$

4. $12 \div 4 = 3$
5. $28 \div 4 = 7$
6. $20 \div 4 = 5$
7. $36 \div 4 = 9$
8. $24 \div 4 = 6$
9. $18 \div 6 = 3$
10. $42 \div 6 = 7$
11. $30 \div 6 = 5$
12. $48 \div 6 = 8$
13. $71 + 34 + 59 + 26 = 190$
14.
$$\begin{array}{r} {}^5\!6\,{}^1\!5 \\ -\ 3\ 9 \\ \hline 2\ 6 \end{array}$$
$$\begin{array}{r} 8\ 4 \\ \times 6\ 2 \\ \hline 1\ 6\ 8 \end{array}$$
15.
$$\begin{array}{r} 2 \\ {}^1\!4\ 8\ 4 \\ \hline 5,2\ 0\ 8 \end{array}$$
16. $5 \times 4 = 20$
    $20 \div 2 = 10$ sq ft
17. $32 \times 25 = 800$ sq in
18. $2 \times 5 = 10$
    $10 \div 2 = 5$ sq yd
19. $24 \div 4 = 6$ gal
20. $40 \div 4 = \$10$

Test 11

1. $2 + 7 + 8 + 3 = 20$
   $20 \div 4 = 5$
2. $4 + 3 + 6 + 7 + 10 = 30$
   $30 \div 5 = 6$
3. $5 + 11 + 14 = 30$
   $30 \div 3 = 10$
4. $7 \times \underline{7} = 49$
5. $7 \times \underline{6} = 42$
6. $7 \times \underline{8} = 56$
7. $8 \times \underline{8} = 64$
8. $8 \times \underline{7} = 56$
9. $8 \times \underline{9} = 72$

10. $63 \div 9 = 7$
11. $48 \div 6 = 8$
12. $40 \div 5 = 8$
13. $21 \div 3 = 7$
14. $32 \div 4 = 8$
15. $80 \div 10 = 8$
16. $7 \times 7 = 49$ sq ft
17. $1 \times 8 = 8$
    $8 \div 2 = 4$ sq mi
18. yes
19. $11 + 16 = 27$ ft
    $27 \div 3 = 9$ yd
20. $3 + 2 + 4 + 3 = 12$
    $12 \div 4 = 3$ pt average
    $12 \div 2 = 6$ qt total

Test 12

1. $28 \div 7 = 4$
2. $63 \div 7 = 9$
3. $56 \div 8 = 7$
4. $16 \div 8 = 2$
5. $14 \div 7 = 2$
6. $35 \div 7 = 5$
7. $24 \div 8 = 3$
8. $72 \div 8 = 9$
9. $48 \div 8 = 6$
10. $42 \div 7 = 6$
11. $21 \div 7 = 3$
12. $40 \div 8 = 5$
13. $49 \div 7 = 7$
14. $32 \div 8 = 4$
15. $64 \div 8 = 8$
16. $56 \div 7 = 8$
17. $7 \times 9 = 63$ sq in
18. $3 \times 16 = 48$ oz
    $48 < 50$
    50 oz bag is more
19. $28 \div 4 = \$7$
20. $1 + 2 + 7 + 10 + 12 + 16 = 48$
    $48 \div 6 = 8$

Unit Test II

1. $36 \div 4 = 9$
2. $24 \div 6 = 4$
3. $32 \div 8 = 4$
4. $49 \div 7 = 7$
5. $56 \div 8 = 7$
6. $20 \div 4 = 5$
7. $42 \div 6 = 7$
8. $60 \div 6 = 10$
9. $16 \div 8 = 2$
10. $35 \div 7 = 5$
11. $48 \div 8 = 6$
12. $7 \div 7 = 1$
13. $12 \div 4 = 3$
14. $36 \div 6 = 6$
15. $70 \div 7 = 10$
16. $12 \div 6 = 2$
17. $8 \div 8 = 1$
18. $56 \div 7 = 8$
19. $28 \div 4 = 7$
20. $42 \div 7 = 6$
21. $24 \div 8 = 3$
22. $32 \div 4 = 8$
23. $80 \div 8 = 10$
24. $21 \div 7 = 3$
25. $18 \div 6 = 3$
26. $4 \div 4 = 1$
27. $14 \div 7 = 2$
28. $72 \div 8 = 9$
29. $40 \div 4 = 10$
30. $64 \div 8 = 8$
31. $6 \div 6 = 1$
32. $48 \div 6 = 8$
33. $40 \div 8 = 5$
34. $24 \div 4 = 6$
35. $63 \div 7 = 9$
36. $30 \div 6 = 5$
37. $28 \div 7 = 4$
38. $8 \div 4 = 2$
39. $54 \div 6 = 9$
40. $16 \div 4 = 4$
41. $35 + 72 + 15 + 48 = 170$

42.
```
  8 9 1
-   3 6
    5 5
```
43.
```
     7 5
   × 5 8
     4
   5 6 0
     2
   3 5 5
 4,3 5 0
```
44. $7 \times 2 = 14$
    $14 \div 2 = 7$ sq yd
45. $5 + 9 + 13 = 27$
    $27 \div 3 = 9$
46. $6 \times 4 = 24$ qt
47. $32 \div 4 = 8$ gal
48. $9 \times 4 = 36$ quarters
49. $20 \div 4 = \$5$
50. $2 \times 16 = 32$ oz

Test 13

1. $8 + 12 = 20$
   $20 \div 2 = 10$
   $10 \times 6 = 60$ sq in
2. $2 \times 7 = 14$
   $14 \div 2 = 7$ sq ft
3. $3 + 9 = 12$
   $12 \div 2 = 6$
   $6 \times 8 = 48$ sq ft
4. $52 \times 36 = 1,872$ sq in
5. $14 \times 20 = 280$
6. $22 \times 30 = 660$
7. $43 \times 30 = 1,290$
8. $51 \times 20 = 1,020$
9. $42 \div 7 = 6$
10. $100 \div 10 = 10$
11. $64 \div 8 = 8$
12. $27 \div 9 = 3$
13.
```
   3 5 2
 + 1 2 6
   4 7 8
```

14.
$$\begin{array}{r} 7 \ ^{10} \ ^1 \\ \cancel{8} \ \cancel{1} \ 1 \\ - \quad 3 \ 4 \ 9 \\ \hline 4 \ 6 \ 2 \end{array}$$

15.
$$\begin{array}{r} 1 \\ 6 \ 0 \ 7 \\ + 7 \ 8 \ 5 \\ \hline 1,3 \ 9 \ 2 \end{array}$$

16. $8 \div 4 = 2$ gal
17. $28 \div 4 = \$7$
18. $27 \div 3 = 9$ yd
19. $50 \div 5 = 10$ ft
20. $\$315 - \$227 = \$88$

Test 14

1. 221,346
2. 3,467,000
3. $6,000,000 + 100,000 +$
   $20,000 + 3,000 + 500$
4. $4,000,000 + 500,000$
5. $6 + 10 = 16$
   $16 \div 2 = 8$
   $8 \times 7 = 56$ sq in
6. $19 \times 25 = 475$ sq ft
7. $6 \times 200 = 1,200$
8. $24 \times 200 = 4,800$
9. $17 \times 100 = 1,700$
10. $32 \times 200 = 6,400$
11. $56 \div 8 = 7$
12. $49 \div 7 = 7$
13. $9 \div 9 = 1$
14. $24 \div 6 = 4$
15. 2
16. $5 \times 16 = 80$ oz
17. $8 \times 2 = 16$ sq in
    $16 \div 2 = 8$ sq in
18. $16 \div 2 = 8$ qt
    $8 \div 4 = 2$ gal

Test 15

1. 7,632,400,000; seven billion,
   six hundred thirty-two million,
   four hundred thousand
2. 555,431,000;
   five hundred fifty-five million,
   four hundred thirty-one thousand
3. 1,635,721,000,000
4. 4,315,021
5. $8 \times 1,000,000,000 +$
   $2 \times 100,000,000 + 5 \times 10,000,000$
6. $3 \times 1,000,000,000,000 +$
   $4 \times 100,000,000,000 +$
   $2 \times 100 + 7 \times 10 + 4 \times 1$
7. $10 \times 60 = 600$
8. $12 \times 40 = 480$
9. $20 \times 200 = 4,000$
10.
$$\begin{array}{r} 1 \ 1 \\ 2,5 \ 4 \ 3 \\ + 8,0 \ 6 \ 7 \\ \hline 10,6 \ 1 \ 0 \end{array}$$

11.
$$\begin{array}{r} ^3 \ ^{15} \ ^1 \\ 6, \ \cancel{4} \ \cancel{6} \ \cancel{0} \\ - \quad 1 \ 9 \ 2 \\ \hline 6, \ 2 \ 6 \ 8 \end{array}$$

12.
$$\begin{array}{r} 1, ^1 2 \ ^1 4 \ 7 \\ 3, 5 \ 9 \ 8 \\ + 6,0 \ 1 \ 3 \\ \hline 10,8 \ 5 \ 8 \end{array}$$

13. $63 \div 9 = 7$
14. $36 \div 6 = 6$
15. $21 \div 7 = 3$
16. $8 \div 8 = 1$
17. $3 + 5 + 11 + 13 = 32$
    $32 \div 4 = 8$
18. $16 \times 12 = 192$ hours

Test 16

1. 3 r.1
$$3\overline{)10}$$
   9
   1

2. 4 r.1
$$6\overline{)25}$$
   24
   1

3. 3 r.3
$$9\overline{)30}$$
   27
   3

4. 7 r.3
$$4\overline{)31}$$
   28
   3

5. 7 r.1
$$2\overline{)15}$$
   14
   1

6. 9 r.2
$$5\overline{)47}$$
   45
   2

7. 2 r.4
$$8\overline{)20}$$
   16
   4

8. 5 r.6
$$7\overline{)41}$$
   35
   6

9. 4 r.5
$$8\overline{)37}$$
   32
   5

10.
```
  2  10 6  16
  3, 0 7 6
- 1, 4 6 7
-----------
  1, 6 0 9
```

11.
```
  4,5 6 14 5 14
- 3, 2  9  8
---------------
  1, 3  5  6
```

12.
```
   6,5 1 2
 + 7,2 8 5
-----------
  13,7 9 7
```

13. $11 \times 700 = 7,700$
14. $12 \times 300 = 3,600$
15. $21 \times 400 = 8,400$
16. 8,310,675,420
17. $18 \div 4 = 4$ r. 2
    4 full cages with 2 left over
18. $45 \times 38 = 1,710$ sq ft

Test 17

1. 1,008; $800 + 200 + 8$
2. 1,008; $800 + 200 + 8$
3. $90 \div 3 = 30$
4. $240 \div 6 = 40$
5. $120 \div 2 = 60$
6. $200 \div 5 = 40$
7. $55 \div 6 = 9$ r. 1
8. $26 \div 8 = 3$ r. 2
9. $64 \div 9 = 7$ r. 1
10. $30 \div 7 = 4$ r. 2
11. $3 \times 16 = 48$ oz
12. $4 \times 4 = 16$ qt
13. $33 \div 3 = 11$ yd
14. $1 \times 2,000 = 2,000$
15. $6 \times 2,000 = 12,000$ lb
16. $3 \times 2,000 = 6,000$ lb
17. $31 \div 4 = 7$ r. 3
    $7 per child with $3 left over
18. 5,400,600,000,005

Test 18

1.
```
    21        2
  2|42      ×21
    40       42
     2
     2
     0
```

2.
```
   10 r.2      9
  9|92       ×10
   90         90
    2        + 2
             92
```

3.
```
   22 r.1      3
  3|67       ×22
   60         66
    7        + 1
    6         67
    1
```

4.
```
    4 r.3      4
  4|19       × 4
   16         16
    3        + 3
             19
```

5.
```
    4 r.1      5
  5|21       × 4
   20         20
    1        + 1
             21
```

6.
```
   60          8
  8|480      ×60
   480        480
    0
```

7.
```
   ¹2 5
   1 7 5
  + 4 5
   1 4 5
```

8.
```
   8
   7
   4
   3
  +2
  24
```

9.
```
   ¹ ¹9 5
  +3 4 5
   4 40
```

10. $4 \times 2{,}000 = 8{,}000$

11. $31 \times 4 = 124$

12. $10 \times 16 = 160$

13. $1 \times 5{,}280 = 15{,}280$

14. $3 \times 5{,}280 = 15{,}840$

15. $5 \times 5{,}5280 = 26{,}400$

16. $4 + 8 + 15 + 25 = 52$

    $52 \div 4 = 13$

17. $25 \div 4 = 6$ r. 1

    6 pages filled, 1 picture left over

18. $5 \times 3 = 15$ ft

Test 19

1.
```
      238
   3|714
    600
    114
     90
     24
     24
      0
```

2.
```
        3
    ×238
      24
       9
       6
     714
```

3.
```
     125 r.3
   5|628
    500
    128
    100
     28
     25
      3
```

4.
$$\begin{array}{r} 5 \\ \times 125 \\ \hline 25 \\ 10 \\ 5 \\ \hline 625 \\ + \quad 3 \\ \hline 628 \end{array}$$

5.
$$\begin{array}{r} 92 \\ 4\overline{)368} \\ 360 \\ \hline 8 \\ 8 \\ \hline 0 \end{array}$$

6.
$$\begin{array}{r} 4 \\ \times 92 \\ \hline 368 \end{array}$$

7.
$$\begin{array}{r} 16\,\mathrm{r.}1 \\ 7\overline{)113} \\ 70 \\ \hline 43 \\ 42 \\ \hline 1 \end{array}$$

8.
$$\begin{array}{r} 7 \\ \times 16 \\ \hline 42 \\ 7 \\ \hline 112 \\ + \quad 1 \\ \hline 113 \end{array}$$

9.
$$\begin{array}{r} 453 \\ \times \quad 46 \\ \hline 231 \\ 408 \\ {}^12\,12 \\ 60 \\ \hline 20,838 \end{array}$$

10.
$$\begin{array}{r} 839 \\ \times \quad 25 \\ \hline 4^14 \\ 055 \\ 1 \\ {}^16\,68 \\ \hline 20,975 \end{array}$$

11.
$$\begin{array}{r} 851 \\ \times \quad 69 \\ \hline {}^14 \\ 7\,259 \\ 3 \\ 4\,806 \\ \hline 58,719 \end{array}$$

12. $40 \times 3 = 120$

13. $2 \times 5,280 = 10,560$

14. $5 \times 16 = 80$

15. $2,495,000,000$

16. parallel

Test 20

1.
$$\begin{array}{r} 179 \\ 5\overline{)895} \\ 500 \\ \hline 395 \\ 350 \\ \hline 45 \\ 45 \\ \hline 0 \end{array}$$

2.
$$\begin{array}{r} 5 \\ \times 179 \\ \hline 895 \end{array}$$

3.
$$\begin{array}{r} 44\,\frac{4}{8} \\ 8\overline{)356} \\ 320 \\ \hline 36 \\ 32 \\ \hline 4 \end{array}$$

4.
$$\begin{array}{r} 8 \\ \times 44 \\ \hline 32 \\ 32 \\ \hline 352 \\ + \quad 4 \\ \hline 356 \end{array}$$

5.
$$204\tfrac{2}{3}$$
3⟌614
600
14
12
2

6.
3
×204
12
6
612
+ 2
614

7.
$$96\tfrac{2}{6}$$
6⟌578
540
38
36
2

8.
6
×96
36
540
576
+ 2
578

9. $300 \div 5 = 60$ mi
10. $3 \times 5,280 = 15,840$ ft
11. $35 \div 7 = 5$ yards
12. base $= 7 \times 3 = 21$ ft
height $= 5 \times 3 = 15$ ft
$21 \times 15 = 315$ sq ft
13. $8 \times 2,000 = 16,000$ lb
$16,000 > 1,600$
8 tons weighs more
14. 4
15. $2,003 - 1,974 = 29$ years

Test 21

1. 30
2. 100
3. 1,000
4. 10
5. 400
6. 5,000
7.
(100)
6⟌(900)

8.
$$144\tfrac{5}{6}$$
6⟌869
600
269
240
29
24
5

9.
326
2⟌652
600
52
40
12
12
0

10.
2
×326
12
4
6
652

11.
$$34\tfrac{1}{7}$$
7⟌239
210
29
28
1

12.
$$\begin{array}{r} 7 \\ \times 34 \\ \hline 28 \\ 21 \\ \hline 238 \\ + 1 \\ \hline 239 \end{array}$$

13. $2 \times 5,280 = 10,560$ ft
14. $5 \times 2,000 = 10,000$ lb
15. $5 \times 500 = 2,500$ mi
16. $13 + 17 = 30$
$30 \div 2 = 15$
$15 \times 11 = 165$ sq ft

Unit Test III

1. $5 + 19 = 24$
$24 \div 2 = 12$
$12 \times 6 = 72$ sq. in
2. $2 + 18 = 20$
$20 \div 2 = 10$
$10 \times 5 = 50$ sq. ft
3. 3,761,800,000
4. 2,413,283,350,000
5. $7 \times 10,000,000 + 5 \times 1,000,000 +$
$1 \times 100,000 + 2 \times 10,000 + 3 \times 1,000$
6. $28 \times 60 = 1,680$
7. $13 \times 400 = 5,200$
8. $56 \times 700 = 39,200$
9.
$$\begin{array}{r} 30 \\ 2\overline{)60} \\ 60 \\ \hline 0 \end{array}$$
10.
$$\begin{array}{r} 8\,r.1 \\ 8\overline{)65} \\ 64 \\ \hline 1 \end{array}$$
11.
$$\begin{array}{r} 9\,r.3 \\ 4\overline{)39} \\ 36 \\ \hline 3 \end{array}$$

12.
$$\begin{array}{r} 223 \\ 3\overline{)669} \\ 600 \\ \hline 69 \\ 60 \\ \hline 9 \\ 9 \\ \hline 0 \end{array}$$
13.
$$\begin{array}{r} 51\,r.2 \\ 5\overline{)257} \\ 250 \\ \hline 7 \\ 5 \\ \hline 2 \end{array}$$
14.
$$\begin{array}{r} 103\,r.5 \\ 8\overline{)829} \\ 800 \\ \hline 29 \\ 24 \\ \hline 5 \end{array}$$
15. 30
16. 200
17. 3,000
18.
$$\begin{array}{r} (100) \\ 5\overline{)(900)} \end{array}$$
19.
$$\begin{array}{r} 181\frac{3}{5} \\ 5\overline{)908} \\ 500 \\ \hline 408 \\ 400 \\ \hline 8 \\ 5 \\ \hline 3 \end{array}$$
20.
$$\begin{array}{r} 216\frac{1}{4} \\ 4\overline{)865} \\ 800 \\ \hline 65 \\ 40 \\ \hline 25 \\ 24 \\ \hline 1 \end{array}$$

21.
$$
\begin{array}{r}
4 \\
\times 216 \\
\hline
24 \\
4 \\
8 \\
\hline
864 \\
+\ 1 \\
\hline
865
\end{array}
$$

22.
$$
78\frac{1}{2}
$$
$$
2\overline{)157}
$$
$$
\begin{array}{r}
140 \\
\hline
17 \\
16 \\
\hline
1
\end{array}
$$

23.
$$
\begin{array}{r}
2 \\
\times 78 \\
\hline
16 \\
14 \\
\hline
156 \\
+\ 1 \\
\hline
157
\end{array}
$$

24. $5 \times 5{,}280 = 26{,}400$ ft

25. $9 \times 2{,}000 = 18{,}000$ lb

26. $568 \div 2 = 284$ people

Test 22

1.
$$
15\frac{7}{34}
$$
$$
34\overline{)517}
$$
$$
\begin{array}{r}
340 \\
\hline
177 \\
170 \\
\hline
7
\end{array}
$$

2. $15 \times 34 = 510$
   $510 + 7 = 517$

3.
$$
20\frac{7}{18}
$$
$$
18\overline{)367}
$$
$$
\begin{array}{r}
360 \\
\hline
7
\end{array}
$$

4. $20 \times 18 = 360$
   $360 + 7 = 367$

5.
$$
15\frac{3}{5}
$$
$$
5\overline{)78}
$$
$$
\begin{array}{r}
50 \\
\hline
28 \\
25 \\
\hline
3
\end{array}
$$

6. $15 \times 5 = 75$
   $75 + 3 = 78$

7.
$$
103\frac{7}{9}
$$
$$
9\overline{)934}
$$
$$
\begin{array}{r}
900 \\
\hline
34 \\
27 \\
\hline
7
\end{array}
$$

8. $103 \times 9 = 927$
   $927 + 7 = 934$

9. $20 + 40 = 60$
   $60 \div 2 = 30$
   $30 \times 30 = 900$ sq ft

10. $12 \times 59 = 708$
    $708 \div 2 = 354$ sq in

11. 455 sq ft

12. $144 \div 12 = 12$

13. $20 \times 12 = 240$

14. $72 \div 12 = 6$

15. $48 \div 4 = \$12$

16. $330 \div 55 = 6$ hours

17. $330 \div 11 = 30$ miles

18. $7 \times 2{,}000 = 14{,}000$ lb
    $140{,}000 > 14{,}000$

Test 23

1.
$$
912\frac{2}{5}
$$
$$
5\overline{)4562}
$$
$$
\begin{array}{r}
4500 \\
\hline
62 \\
50 \\
\hline
12 \\
10 \\
\hline
2
\end{array}
$$

2.
```
        5
     ×912
       10
        5
      ¹45
    4,560 + 2 = 4,562
```

3.
```
       213 2/7
    7⟌1493
       1400
         93
         70
         23
         21
          2
```

4.
```
        7
     ×213
       21
        7
       14
     1491
    +   2
    1,493
```

5.
```
        4 30/82
    82⟌358
       328
        30
```

6.
```
       4
     ×82
       8
      32
     328
    +30
     358
```

7.
```
        40
    21⟌840
       840
         0
```

8.
```
      21
    ×40
    840
```

9.
```
     4,137
    ×   13
         2
    12¹391
     4137
    53,781
```

10.
```
      2,428
    ×    75
    ¹¹2¹14
      0000
      215
    14846
    182,100
```

11.
```
       7,801
     ×    36
          4
      ¹42806
          2
      21403
      280,836
```

12. 396 ÷ 12 = 33 ft
13. 3 parallel lines
14. 80 ÷ 16 = 5 lb
     5 < 10; yes
15. 2,272 × 4 = 9,088 qt   *568 qt*

Test 24

1.
```
         108 24/25
    25⟌2724
       2500
        224
        200
         24
```

2.
```
        25
     ×108
       14
      160
      25
     2700
    +  24
    2724 +1 = 2725
```

3.
$$\begin{array}{r} 28\frac{66}{81} \\ 81\overline{)2334} \\ 1620 \\ \hline 714 \\ 648 \\ \hline 66 \end{array}$$

4.
$$\begin{array}{r} 81 \\ \times\ 28 \\ \hline 1\ 648 \\ 1\ 62 \\ \hline 1\ 1 \\ 2\ 268 \\ +\ \ 66 \\ \hline 2,334 \end{array}$$

5.
$$\begin{array}{r} 1445\frac{3}{6} \\ 6\overline{)8673} \\ 6000 \\ \hline 2673 \\ 2400 \\ \hline 273 \\ 240 \\ \hline 33 \\ 30 \\ \hline 3 \end{array}$$

6.
$$\begin{array}{r} 6 \\ \times 1445 \\ \hline 30 \\ 24 \\ 24 \\ 6 \\ \hline 8670 \\ +\ \ \ 3 \\ \hline 8673 \end{array}$$

7.
$$\begin{array}{r} 31 \\ 12\overline{)372} \\ 360 \\ \hline 12 \\ 12 \\ \hline 0 \end{array}$$

8.
$$\begin{array}{r} 12 \\ \times 31 \\ \hline 12 \\ 36 \\ \hline 372 \end{array}$$

9.
$$\begin{array}{r} 345 \\ \times 1472 \\ \hline {}^16\ 1 \\ {}^1{}_2{}^12\ 380 \\ 1\ 85 \\ {}^11\ 2 \\ 2\ 60 \\ 345 \\ \hline 507,840 \end{array}$$

10.
$$\begin{array}{r} 837 \\ \times 4754 \\ \hline {}^23\ 12 \\ 228 \\ {}^14\ 13 \\ 0\ 55 \\ {}^15\ 2\ 4 \\ 6\ 19 \\ 3\ 12 \\ 2\ 28 \\ \hline 3,979,098 \end{array}$$

11.
$$\begin{array}{r} 632 \\ \times 3581 \\ \hline 632 \\ {}^24{}^12\ 1 \\ 8\ 46 \\ {}^13\ 1\ 1 \\ 0\ 50 \\ 1\ 896 \\ \hline 2,263,192 \end{array}$$

12. $13 + 19 = 32$
    $32 \div 2 = 16$
    $16 \times 41 = 656$ sq ft

13. $34 \div 5 = 6$ r. 4
    6 full cars and 1 part car; 7 cars

14. ‖

Test 25

1.
$$1153 \frac{19}{29}$$
$$29\overline{)33456}$$
$$\underline{29000}$$
$$4456$$
$$\underline{2900}$$
$$1556$$
$$\underline{1450}$$
$$106$$
$$\underline{87}$$
$$19$$

2.
```
      29
 × 1153
      2
  1167
  145
  129
   29
 33437
+    19
 33,456
```

3.
$$2081 \frac{95}{180}$$
$$180\overline{)374675}$$
$$\underline{360000}$$
$$14675$$
$$\underline{14400}$$
$$275$$
$$\underline{180}$$
$$95$$

4.
```
     180
 ×  2081
  6 180
 1684
 2
 374580
+     95
 374,675
```

5.
$$656$$
$$7\overline{)4592}$$
$$\underline{4200}$$
$$392$$
$$\underline{350}$$
$$42$$
$$\underline{42}$$
$$0$$

6.
```
      7
 × 656
     42
    35
   42
  4,592
```

7.
$$108 \frac{11}{15}$$
$$15\overline{)1631}$$
$$\underline{1500}$$
$$131$$
$$\underline{120}$$
$$11$$

8.
```
     15
 × 108
     4
  1580
  1620
+   11
  1631
```

9. $336 \div 12 = 28$ in
10. $37,500 \div 250 = 150$ ft
11. $323 \div 19 = 17$ in
12. $56 \times 16 = 896$ oz
13. $270 \div 6 = 45$ mph
14. $780 + 4,670 + 550 = 6,000$
    $6,000 \div 3 = 2,000$ lb
    $2,000$ lb $= 1$ ton

Test 26

1. $24 \times 12 \times 6 = 1,728$ cu ft
2. $33 \times 45 \times 10 = 14,850$ cu in

3.
```
        5474
    5 ) 27370
      25000
       2370
       2000
        370
        350
         20
         20
          0
```

4.
```
         5
    × 5 4 7 4
        20
        35
       2 0
     2 5
    27,370
```

5.
```
      2570 21/32
   32 ) 82261
      64000
      18261
      16000
       2261
       2240
         21
```

6.
```
        32
     × 2 5 7 0
       2 1
   ⁴1 1 1 4
      5 0
     6 4
     8 2 2 4 0
     +    2 1
     82,261
```

7. $28 \div 2 = 14$
8. $44 \times 4 = 176$
9. $48 \div 16 = 3$
10. $6 \times 2,000 = 12,000$
11. $100 \div 4 = 25$
12. $60 \div 12 = 5$
13. $20 \times 15 \times 6 = 1,800$ cu ft
14. $1,800 \times 7 = 12,600$ gal
15. $12,600 \times 8 = 100,800$ lb

Test 27

1. $21 \div 7 = 3$
   $3 \times 4 = 12$
2. $9 \div 3 = 3$
   $3 \times 2 = 6$
3. $36 \div 9 = 4$
   $4 \times 1 = 4$
4. $75 \div 5 = 15$
   $15 \times 3 = 45$
5. $18 \div 6 = 3$
   $3 \times 5 = 15$
6. $28 \div 2 = 14$
   $14 \times 1 = 14$

7.
```
      9715 2/3
   3 ) 29147
      27000
       2147
       2100
         47
         30
         17
         15
          2
```

8.
```
         3
     × 9 7 1 5
        15
         3
        21
        27
     29145
     +    2
     29147
```

9.
```
      1011 30/46
   46 ) 46536
      46000
        536
        460
         76
         46
         30
```

10.
```
      46
   × 1011
    ¹1 46
      46
    4 6
   46506
   +   30
   46,536
```

11. $36 \div 3 = 12$

12. $24 \div 4 = 6$

13. $8 \times 5,280 = 42,240$

14. $32 \div 4 = 8$

$8 \times 1 = 8$ flowers

15. $100 \times 100 \times 90 = 900,000$ cu in

Test 28

1. 14
2. 72
3. 230
4. 99
5. XLI
6. LXXXV
7. CCCXXXIII
8. XXIX
9. $12 \div 2 = 6$

$6 \times 1 = 6$

10. $64 \div 8 = 8$

$8 \times 3 = 24$

11. $15 \div 5 = 3$

$3 \times 2 = 6$

12. $11 + 25 = 36$

$36 \div 2 = 18$

$18 \times 13 = 234$ sq ft

13.
```
      5373
   5 )26865
     25000
      1865
      1500
       365
       350
        15
        15
         0
```

14.
```
       5
   × 5373
      15
      35
     1 5
    2 5
   26,865
```

15.
```
        2284  140
   216 )493484 216
       432000
        61484
        43200
        18284
        17280
         1004
          864
          140
```

16.
```
       216
   × 2 284
        24
      ¹8 4
    ¹²6 48
       8
       1 2
      4 2
      1 2
     42
   49 3 3 44
   +    140
   49 3,4 84
```

17. $16 + 43 + 58 + 91 = 208$

$208 \div 4 = 52$

18. $24 \div 6 = 4$

$4 \times 1 = 4$ horses

Test 29

1. $\dfrac{5}{6}$

2. $\dfrac{1}{3}$

3.

4.

5. 26
6. 43
7. 165
8. 192
9. XLVII
10. XVIII
11. CCXIX
12. CLIV

13.
$$283\dfrac{20}{31}$$

```
       283 20/31
   31│8793
      6200
      2593
      2480
       113
        93
        20
```

14.
```
        31
      ×283
      ¹193
      248
      62
      8773
    +  20
      8,793
```

15.
```
      1417 7/14
   14│19845
      14000
       5845
       5600
        245
        140
        105
         98
          7
```

16.
```
         14
      ×1417
        128
       114
        46
       14
      198¹38
    +     7
      19,845
```

17. $20 \times 9 = 180$
$180 \div 2 = 90$ sq in

18. $165 \div 55 = 3$ hr

Test 30

1. 2,200
2. 525
3. 750
4. 929
5. LVIII
6. DXX
7. MMMDCC
8. MCMLXV

9. $\dfrac{1}{5}$

10. $\dfrac{3}{3}$

11.
$$127 \frac{5}{64}$$

$$64 \overline{\smash{)}8133}$$
$$\underline{6400}$$
$$1733$$
$$\underline{1280}$$
$$453$$
$$\underline{448}$$
$$5$$

12.
$$\begin{array}{r} 64 \\ \times 127 \\ \hline {}^{1}4\,4\,8 \\ {}^{1}1\,2\,8 \\ 6\,4 \\ \hline 8\,1\,{}^{1}2\,8 \\ +\quad 5 \\ \hline 8,1\,3\,3 \end{array}$$

13.
$$1286$$

$$500 \overline{\smash{)}643000}$$
$$\underline{500000}$$
$$143000$$
$$\underline{100000}$$
$$43000$$
$$\underline{40000}$$
$$3000$$
$$\underline{3000}$$
$$0$$

14.
$$\begin{array}{r} 500 \\ \times 1286 \\ \hline 3000 \\ 40000 \\ 1000 \\ 500 \\ \hline 643,000 \end{array}$$

15. $7 + 9 = 16$
$16 \div 2 = 8$
$8 \times 8 = 64$ sq ft

16. 600

17. $135 \div 15 = 9$ hours

18. $25 \div 5 = 5$
$5 \times 1 = 5$ lb spoiled
$25 - 5 = 20$ lb left

Unit Test IV

1.
$$43 \frac{9}{13}$$

$$13 \overline{\smash{)}568}$$
$$\underline{520}$$
$$48$$
$$\underline{39}$$
$$9$$

2.
$$\begin{array}{r} 13 \\ \times 43 \\ \hline 139 \\ 42 \\ \hline 5{}^{1}59 \\ +\quad 9 \\ \hline 568 \end{array}$$

3.
$$32 \frac{11}{30}$$

$$30 \overline{\smash{)}971}$$
$$\underline{900}$$
$$71$$
$$\underline{60}$$
$$11$$

4.
$$\begin{array}{r} 30 \\ \times 32 \\ \hline 60 \\ 90 \\ \hline 960 \\ +11 \\ \hline 971 \end{array}$$

5.
$$9030 \frac{4}{5}$$

$$5 \overline{\smash{)}45154}$$
$$\underline{45000}$$
$$154$$
$$\underline{150}$$
$$4$$

6.
$$\begin{array}{r} 5 \\ \times 9030 \\ \hline 15 \\ 45 \\ \hline 45150 \\ +\quad 4 \\ \hline 45154 \end{array}$$

Final Test

7.
$$2872 \frac{4}{24}$$
24 $\overline{)68932}$
$\underline{48000}$
20932
$\underline{19200}$
1732
$\underline{1680}$
52
$\underline{48}$
4

1.
$$20$$
4 $\overline{)80}$
$\underline{80}$
0

2.
$$7 \text{ r.}4$$
7 $\overline{)53}$
$\underline{49}$
4

8.
$$24$$
$\times 2872$
48
1^12
48
3
^{162}
$\underline{48}$
68928
$\underline{+\quad 4}$
68,932

3.
$$81$$
8 $\overline{)648}$
$\underline{640}$
8
$\underline{8}$
0

4.
$$79 \text{ r.}1$$
5 $\overline{)396}$
$\underline{350}$
46
$\underline{45}$
1

9. $16 \times 8 \times 7 = 896$ cu ft
10. $25 \times 21 \times 20 = 10,500$ cu in
11. $32 \div 2 = 16$
     $16 \times 1 = 16$
12. $64 \div 8 = 8$
     $8 \times 5 = 40$
13. $42 \div 6 = 7$
     $7 \times 5 = 35$
14. $\frac{4}{5}$
15. $\frac{1}{3}$
16. $8 \times 12 = 96$ in
17. 99
18. MMCDLIII

5.
$$25 \frac{6}{25}$$
25 $\overline{)631}$
$\underline{500}$
131
$\underline{125}$
6

6.
$$25$$
$\times 25$
125
10
$\underline{40}$
$6^12 5$
$\underline{+\quad 6}$
631

7.
$$21 \frac{13}{16}$$
16 $\overline{)349}$
$\underline{320}$
29
$\underline{16}$
13

8.
```
      16
    × 21
      16
      32
     336
    + 13
     349
```

9.
$$5076 \frac{2}{6}$$
```
6 ) 30458
    30000
      458
      420
       38
       36
        2
```

10.
```
         6
    × 5 076
        36
        42
       3 0
     30 456
    +     2
    30,458
```

11.
$$686 \frac{23}{84}$$
```
84 ) 57647
     50400
      7247
      6720
       527
       504
        23
```

12.
```
        84
    ×  6 86
      ¹4 2
     ¹6 3 84
    ¹4 2 4 2
      8 4
     5 7 6 24
    +     23
    5 7,6 47
```

13. $3+7 = 10$
$10 \div 2 = 5$
$5 \times 5 = 25$ sq ft

14. $4 \times 9 = 36$
$36 \div 2 = 18$ sq ft

15. $10 \times 25 = 250$ sq in

16. $33 \times 15 = 495$ sq ft

17. $45 \times 29 \times 12 = 15{,}660$ cu ft

18. $27 \div 3 = 9$

19. $40 \div 2 = 20$

20. $20 \div 4 = 5$

21. $5 \times 4 = 20$

22. $4 \times 16 = 64$

23. $1 \times 5{,}280 = 5{,}280$

24. $5 \times 2{,}000 = 10{,}000$

25. $36 \div 12 = 3$

26. $40 \times 12 = 480$

27. 50

28. 4,000

29. 500

30. $12 \div 3 = 4$
$4 \times 1 = 4$

31. $21 \div 7 = 3$
$3 \times 3 = 9$

32. $32 \div 8 = 4$
$4 \times 5 = 20$

33. $\dfrac{3}{5}$

34. $\dfrac{2}{3}$

35. 2,543,900,000

36. $5+12+13+21+24 = 75$
$75 \div 5 = 15$

37. 2,158

38. MCMLXXV

Word Problems Lesson 15

1. area of garden is $11 \times 13 = 143$ sq ft

    $10 \times 10 = 100$ sq ft needed for one packet

    $2 \times 100 = 200$ sq ft needed for 2 packets

    $200 > 143$, so garden does not have enough space.

2. Use a drawing to show the girls' travels. The distances don't have to be exact.

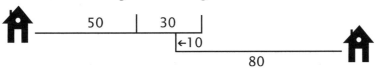

    $50 + 30 = 80$ to turn around

    $80 + 10 = 90$ back-track to restaurant

    $90 + 80 = 170$ miles is total distance driven

3. $\$50 + \$20 = \$70$ what they left with

    plus $\$10$ to each

    $\$8 + \$15.65 + \$10 = \$33.65$ what they spent

    (gifts are $\$5 + \$5$)

    $\$70 - \$33.65 = \$36.35$ left

Word Problems Lesson 21

1. $3 \times 75 = 225$ pieces of candy

    $225 \div 7 = 32$ r. 1

    Scott will have one piece left

2. $5 \times \$3 = \$15$ for baby-sitting

    $3 \times \$4 = \$12$ for garden work

    $\$15 + \$12 = \$27$ she has

    $\$35 - \$27 = \$8$ more needed to buy the game

3. $25 \div 5 = 5$ acorns in a group

    $16 \div 4 = 4$ seed pods in a group

    $8 \div 2 = 4$ feathers in a group

    $5 + 4 + 4 = 13$ items given to Mom

Word Problems Lesson 27

1. $65 \div 13 = 5$ bags per student

   $5 \times 15 = 75$ nuts per student

2. You may want to draw this one.

   biggest move is 5, so two times biggest move is 10.

   $3 + 5 + 1 = 9$

   $9 - 6 = 3$

   $3 + 10 = 13$, so he is 13 spaces from the beginning.

3. $3 \times 11 = 33$

   $V = 33 \times$ missing measure

   V is 330

   $330 \div 33 = 10$, which is the missing measure of the side of the box top

   $10 \times 11 = 110$ sq ft which is the area of the top of the box.

4. $\$360 \div 6 = \$60$ for each child

   $\$60 + \$20 = \$80$ Kate's money

   $\$80 \div 16 = \$5$ cost of each gift

Symbols & Tables

MONEY

1 nickel = 5 cents (5¢)

1 dime = 10 cents (10¢)

1 quarter = 25 cents (25¢)

1 dollar = 100 cents (100¢ or $1.00)

1 dollar = 4 quarters

MEASUREMENT

3 teaspoons (tsp) = 1 tablespoon (Tbsp)

2 pints (pt) = 1 quart (qt)

8 pints = 1 gallon (gal)

4 quarts = 1 gallon

12 inches (in) = 1 foot (ft)

3 feet = 1 yard (yd)

5,280 feet = 1 mile (mi)

16 ounces (oz) = 1 pound (lb)

2,000 pounds = 1 ton

60 seconds = 1 minute

60 minutes = 1 hour

7 days = 1 week

365 days = 1 year

52 weeks = 1 year

12 months = 1 year

1 dozen = 12

1 cubic foot of water is about 7 gallons

1 gallon of water weighs about 8 pounds

PLACE-VALUE NOTATION

$31,452 = 30,000 + 1,000 + 400 + 50 + 2$

EXPANDED NOTATION

$1,452 = 1 \times 1,000 + 4 \times 100 + 5 \times 10 + 2 \times 1$

SYMBOLS

=   equals

≈   approximately equal to

+   plus

−   minus

x   times

•   times

( )( )   times

¢   cents

$ dollars

'   foot

"   inch

<   less than

>   greater than

|| parallel

└   right angle

⊥   perpendicular

$4 \div 2$ 4 divided by 2

$2\overline{)4}$ 4 divided by 2

$\dfrac{4}{2}$ 4 divided by 2

AREA AND VOLUME

rectangle $A = bh$ (base times height)

parallelogram $A = bh$

triangle $\qquad A = \dfrac{bh}{2}$

trapezoid $\qquad A = \dfrac{b_1 + b_2}{2} \times h$

rectangular solid $\qquad V = Bh$

$\qquad\qquad$ (area of base times height)

LABELS FOR PARTS OF PROBLEMS

Addition

$$\begin{array}{rl}
25 & \text{addend} \\
+16 & \text{addend} \\
\hline
41 & \text{sum}
\end{array}$$

Multiplication

$$\begin{array}{rl}
33 & \text{multiplicand} \\
\times\ 5 & \text{multiplier} \\
\hline
165 & \text{product}
\end{array}$$

Subtraction

$$\begin{array}{rl}
45 & \text{minuend} \\
-\ 22 & \text{subtrahand} \\
\hline
23 & \text{difference}
\end{array}$$

Division

$$\text{divisor } 2\overline{)4}\ \begin{array}{l}2 \quad \text{quotient}\\ \quad \text{dividend}\end{array}$$

218

Glossary

A

Area - the number of square units in a rectangle or other two-dimensional figure

Average - the result of adding a series of numbers and dividing by the number of items in the series

Base - the top or bottom side of a shape

B-C

Borrowing - see Regrouping

Carrying - see Regrouping

Commutative property - the order of factors in a multiplication problem may be changed without changing the product. The commutative property also applies to addition.

Cube - a three-dimensional figure with each side the same length

Cubic units - the result of multiplying three dimensions. Answers to volume problems are in cubic units.

D

Denominator - the bottom number in a fraction. It tells how many total parts there are in the whole.

Dimension - the length of one of the sides of a rectangle or other shape

Dividend - the number being divided in a division problem

Divisor - the number that is being divided by in a division problem

E

Equation - a number sentence in which the value of one side is equal to the value of the other side

Estimation - used to get an approximate value of an answer

Even number - a number that ends in 0, 2, 4, 6, or 8. Even numbers are multiples of two.

Expanded notion - a way of writing numbers in which each amount is multiplied by its place value

F-G

Factors - the two sides of a rectangle or the numbers multiplied in a multiplication problem

Fraction - one number written over another to show part of a whole. A fraction can also indicate division.

H-K

Height - the length of a line from the top to the bottom of a shape. It forms a right angle with the base.

L-O

Numerator - the top number in a fraction. It tells how many of the parts of a whole have been chosen.

P

Partial product - the result of multiplying by one digit of a multiple-digit problem

Parallel lines - two straight lines in the same plane that never cross or touch

Parallelogram - a shape with two pairs of parallel lines. A rectangle is a special kind of parallelogram.

Perpendicular lines - two straight lines that form a right angle where they meet

Place value - the position of a number that tells what value it is assigned

Place-value notation - a way of writing numbers to emphasize the place value of each part

Plane - a flat surface

Product - the answer to a multiplication problem

Q-R

Quadrilateral - a four-sided figure

Quotient - the answer to a division problem

Rectangle - a shape with four "square corners" or right angles

Rectangular solid - a three-dimensional shape with each side or face shaped like a rectangle

Regrouping - moving numbers from one place value to another in order to solve a problem. Also called "carrying" in addition and multiplication and "borrowing" in subtraction.

Right angle - a square corner (90° angle)

Roman numerals - a numbering system employed by the Roman Empire that uses letters to represent the numbers

Rounding - writing a number as its closest ten, hundred, etc. in order to estimate

S-T

Square - a rectangle with all four sides the same length

Square units - the result of multiplying two dimensions. Answers to area problems are in square units.

Trapezoid - a four-sided shape or figure with two parallel sides

Triangle - a shape with three sides

U-V

Units - the first place value in the decimal system—also, starting from the right, the first three digits in a large number. The word units can also name measurements. Inches and feet are units of measure.

Volume - the number of cubic units in a three-dimensional shape

Master Index for General Math

This index lists the levels at which main topics are presented in the instruction manuals for Primer through Zeta. For more detail, see the description of each level at www.mathusee.com. (Many of these topics are also reviewed in subsequent student books.)

Addition
    facts .... Primer, Alpha
    multiple digit .... Beta
Additive inverse .... Epsilon
Angles .... Zeta
Area
    circle .... Epsilon, Zeta
    parallelogram .... Delta
    rectangle .... Primer, Gamma, Delta
    square .... Primer, Gamma, Delta
    trapezoid................................. Delta
    triangle................................... Delta
Average .... Delta
Circle
    area............................. Epsilon, Zeta
    circumference................. Epsilon, Zeta
    recognition .... Primer, Alpha
Circumference .... Epsilon, Zeta
Common factors .... Epsilon
Composite numbers...................... Gamma
Congruent .... Zeta
Counting...................... Primer, Alpha
Decimals
    add and subtract .... Zeta
    change to percent............ Epsilon, Zeta
    divide .... Zeta
    from a fraction .... Epsilon, Zeta
    multiply .... Zeta
Division
    facts .... Delta multiple
    digit .... Delta
    Estimation .... Beta, Gamma, Delta
Expanded notation.................... Delta, Zeta
Exponential notation .... Zeta
Exponents...................................... Zeta
Factors .... Gamma, Delta, Epsilon
Fractions
    add and subtract .... Epsilon
    compare .... Epsilon
    divide .... Epsilon
    equivalent................. Gamma, Epsilon
    fractional remainders....... Delta, Epsilon
    mixed numbers .... Epsilon
    multiply.................................. Epsilon

    of a number................... Delta, Epsilon
    of one................... Delta, Epsilon
    rational numbers .... Zeta
    reduce .... Epsilon
    to decimals.................... Epsilon, Zeta
    to percents .... Epsilon, Zeta
Geometry
    angles .... Zeta
    area..
    .....Primer, Gamma, Delta, Epsilon, Zeta
    circumference.................. Epsilon, Zeta
    perimeter .... Beta
    plane .... Zeta
    points, lines, rays .... Zeta
    shape recognition........... Primer, Alpha
Graphs
    bar and line .... Beta
    pie... Zeta
Inequalities .... Beta
Linear measure .... Beta, Epsilon
Mean, median, mode......................... Zeta
Measurement equivalents........ Beta, Gamma
Metric measurement .... Zeta
Mixed numbers .... Epsilon
Money.............................. Beta, Gamma
Multiplication
    facts .... Gamma
    multiple digit .... Gamma
Multiplicative inverse .... Epsilon
Perpendicular lines............................ Delta
Pi .... Epsilon, Zeta
Pie graph .... Zeta
Place value...... Primer, Alpha, Beta, Gamma, Delta, Zeta
Prime factorization.......................... Epsilon
Prime numbers .... Gamma
Probability .... Zeta
Rational numbers............................. Zeta
Reciprocal..................................... Epsilon
Rectangle
    area................ Primer, Gamma, Delta
    perimeter .... Beta
    recognition .... Primer, Alpha
Rectangular solid .... Delta

Regrouping.............................Beta, Gamma
Roman numerals..................................Delta
Rounding....................Beta, Gamma, Delta
Sequencing.. Beta
Skip counting
    2, 5, 10....................Primer, Alpha, Beta
    all... Gamma
Solve for unknown ...Primer, Alpha, Gamma,
    Delta, Epsilon, Zeta
Square
    area.................. Primer, Gamma, Delta
    perimeter .... Beta
    recognition...................... Primer, Alpha
Subtraction
    facts .... Alpha
    introduction .... Primer, Alpha
    multiple digit................................ Beta
Tally marks .... Primer, Beta
Telling time ....Primer, Alpha, Beta
Thermometers and Gauges .... Beta
Trapezoid ....Delta
Triangle
    area...Delta
    perimeter .... Beta
    recognition...................... Primer, Alpha
Volume of rectangular solid ....Delta

Delta Index

Addition
    column ....Student 9D
    four digit ....Student 15D
    three digit ....Student 12D
    regrouping ....Student 4D
Area... 1
    parallelogram 7
    rectangle ....Student 1D
    square ....Student 1D
    trapezoid 13
    triangle.................................. 9
    .... 7, 9, 13
Bme.................................... 15
Borrowing ....see regrouping
Carrying..........................see regrouping
Commutative property 1, 3
Cube..................................... 26
Denominator.............................. 27
Dimension 1
Dividend 4
Division
    by 1 2
    by 2 2
    by 3 4
    by 4 10
    by 5 4
    by 6 8
    by 7 12
    by 8 12
    by 9 6
    by 10 3
    fractional remainders.................... 20
    multiple digit.............. 17, 18, 19, 20,
    22, 23, 24, 25
    single digit with remainder 16
    symbols.................................2, 3
Divisor...................................... 4
Dollar ....Student 10D
Estimation 21
Equation 1
Expanded notation.............................. 15
Factors 1, 2
Foot .... Student 1D, 3D, 18D, 22D
    fractional remainders.................... 20
    of a number............................ 27
    of one................................. 29
Gallon..............................Student 10D
Height 7, 9, 13

Inch .... Student 1D, 22D
Measurement
    foot .... Student 1D, 3D, 18D, 22D
    gallon..............................Student 10D
    inch............................ Student 1D, 22D
    mile................................Student 18D
    ounce..............................Student 12D
    pint ....Student 3D
    pound .... Student 12D, 17D
    quart .... Student 3D, 10D
    ton .... Student 17D
    yard...............................Student 3D
Mental math 12, 18, 24
Mile ....Student 18D
Millions................................... 14
Money................................Student 10D
Multiplication
    by multiples of 10 ....Student 13D
    by multiples of 100 ....Student 14D
    facts review ....Student 2, 3
    regrouping ....Student 8D
    review (solving for unknown) . Student 1
    upside down................17, Student 24D
Numerator 27
Ounce...............................Student 12D
Parallel.................................... 5
Parallelogram............................... 7
Partial product 19
Perpendicular............................... 5
Pint ....Student 3D
Place value................................ 14
Place value notation .... 14, Student 8D
Plane 5
Pound .... Student 12D, 17D
Product...................................1, 2
Quadrilateral............................... 13
Quart............................. Student 3D, 10D
Quarter.............................Student 10D
Quotient 4
Rectangle.....................1, 3, Student 1D
Regrouping
    addition...........................Student 4D
    multiplication ....Student 8D
    subtraction ....Student 6D
Right angle 1, 5
Roman numerals 28, 30
Rounding................................... 21
Solving for the unknown 1, 2

Square .... 1, Student 1D
Subtraction
    four digit .... Student 15E
    regrouping .... Student 6D
    three digit .... Student 12E
Thousands 14
Ton .... Student 17D
Trapezoid 13
Triangle 9
Trillions 15
Units...................................... 14
Volume 26
Word problems
    multistep............................ 15, 21, 27
    tips for solving 1, 15
Yard............................... Student 3D